Wood in Construction

Wood in Construction

How to Avoid Costly Mistakes

Jim Coulson

WILEY-BLACKWELL

A John Wiley & Sons, Ltd., Publication

This edition first published 2012 © 2012 by John Wiley & Sons, Ltd

Wiley-Blackwell is an imprint of John Wiley & Sons, formed by the merger of Wiley's global Scientific, Technical and Medical business with Blackwell Publishing.

Registered Office
John Wiley & Sons, Ltd, The Atrium, Southern Gate, Chichester, West Sussex, PO19 8SQ, UK

Editorial Offices
9600 Garsington Road, Oxford, OX4 2DQ, UK
The Atrium, Southern Gate, Chichester, West Sussex, PO19 8SQ, UK
2121 State Avenue, Ames, Iowa 50014-8300, USA

For details of our global editorial offices, for customer services and for information about how to apply for permission to reuse the copyright material in this book please see our website at www.wiley.com/wiley-blackwell.

Library of Congress Cataloging-in-Publication Data

Coulson, Jim (James C.)
   Wood in construction / Jim Coulson.
      p.   cm.
   Includes bibliographical references and index.
   ISBN 978-0-470-65777-5 (pbk. : alk. paper)
1. Building, Wooden.   2. Building, Wooden–Great Britain.   3. Timber.   4. Timber–Great Britain.   5. Wood–Preservation.   6. Wood–Preservation–Great Britain.   I. Title.
   TA419.C68 2012
   694–dc23
                                                          2011035190

A catalogue record for this book is available from the British Library.

Wiley also publishes its books in a variety of electronic formats. Some content that appears in print may not be available in electronic books.

Set in 10/12pt Minion by SPi Publisher Services, Pondicherry, India
Printed and bound in Singapore by Markono Print Media Pte Ltd

1   2012

Contents

Preface   x

1   Wood as a Material   **1**
    1.1   Cellulose   2
    1.2   Grain   4
    1.3   Dimensional change in wood   8
    1.4   Heartwood and sapwood   10
    1.5   Natural durability   13
    1.6   Rays   14
    1.7   Radial and tangential directions   15
    1.8   Permeability in timber   16
    1.9   Chapter summary   16

2   More on Timber: Some Stuff About the UK Timber Trade   **18**
    2.1   'Wood' or 'timber'?   18
    2.2   Wood species and timber trading   19
    2.3   Softwoods and hardwoods   19
    2.4   Some more detail on cell structure   23
    2.5   Trade names and scientific names   27
    2.6   A cautionary tale about timber names   28
    2.7   Growth rings   31
    2.8   Earlywood and latewood   32
    2.9   Rate of growth in softwoods   33
    2.10 Rate of growth in hardwoods   34
    2.11 Chapter summary   37

3   Water in Wood: Moisture Content and the Drying of Timber   **39**
    3.1   The definition of moisture content   39
    3.2   Moisture meters   40
    3.3   'Wet' or 'dry'? In-service moisture contents and 'EMC'   43
    3.4   EMC   44
    3.5   Specification of desired moisture content   44
    3.6   Fibre saturation point   46
    3.7   Shrinkage   47
    3.8   Movement   48
    3.9   Kiln drying   49
    3.10 Air drying   53

3.11	Timescales for drying timber	54
3.12	Chapter summary	55

4   Specifying Timber – for Indoor or Outdoor Uses   57
4.1	British and European standards	57
4.2	Durability and treatability of different wood species	58
4.3	Use classes	59
4.4	Examples of timbers employed in different use classes	60
4.5	Hazard and risk – and their relative importance	60
4.6	Use Class 1 – examples	61
4.7	Use Class 2 – examples	64
4.8	Use Class 3 – examples	66
4.9	Use Class 4 – examples	70
4.10	Use Class 5 – examples	76
4.11	Chapter summary	78

5   The Quality of Timber: Grading for Appearance   79
5.1	The need for grading	80
5.2	'Quality' or 'grade'?	81
5.3	Quality	82
5.4	Grade	82
5.5	The different types of grading	83
5.6	Appearance grading	83
5.7	Appearance grading: based on defects	83
5.8	Scandinavian grades	87
5.9	Unsorted, fifths and sixths	88
5.10	Russian softwood qualities	89
5.11	European appearance grading	90
5.12	North American softwood appearance grades	90
5.13	Clears, merchantable and commons	92
5.14	A comparison of Scandinavian grades and North American grades	93
5.15	Appearance grading: based on 'cuttings'	93
5.16	The NHLA grades	94
5.17	FAS, selects and commons	96
5.18	Selects	96
5.19	'F1F'	97
5.20	Prime and Comsel grades	97
5.21	Malaysian grades	98
5.22	Prime, select and standard	98
5.23	'PHND', 'BHND' or 'sound'	99
5.24	Rules are made to be bent! (within reason)	99
5.25	Shipper's usual	99
5.26	BS EN 942: the quality of timber in joinery	100
5.27	J classes	100

| | 5.28 | 'Exposed face' | 101 |
| | 5.29 | Chapter summary | 102 |

6	**Strength Grading and Strength Classes of Timber**	**104**
	6.1	Appearance versus strength	106
	6.2	Visual strength grades	107
	6.3	GS and SS strength grades	107
	6.4	Strength classes for softwoods	108
	6.5	BS EN 1912	108
	6.6	SC3, SC4: C16 and C24	109
	6.7	Machine grading	110
	6.8	Other strength grades: Europe and North America	112
	6.9	Select structural, No. 1 and No. 2 structural and stud grades	112
	6.10	TR26	113
	6.11	Specifying the strength class or the wood species: some things to think about	114
	6.12	Hardwood strength grades	115
	6.13	Tropical hardwoods	116
	6.14	Temperate hardwoods	116
	6.15	The 'Size effect'	117
	6.16	Hardwood strength classes	118
	6.17	The marking of strength graded timber	119
	6.18	Chapter summary	120

7	**Wood Preservatives and Wood Finishes**	**122**
	7.1	Treat the timber last!	122
	7.2	Wood preservative types	123
	7.3	'Old' and 'new' types of treatments	124
	7.4	The basic methods of timber treatment	124
	7.5	Low pressure treatment	125
	7.6	High pressure treatment	126
	7.7	Preservative chemicals	126
	7.8	CCA preservatives	126
	7.9	The 'environmentally-friendly' preservatives	127
	7.10	'Tanalised' timber	128
	7.11	Organic compounds	128
	7.12	'Treated' timber	129
	7.13	'Wood finishes'	129
	7.14	Wood in exterior uses	130
	7.15	Exterior finishes	131
	7.16	Varnish – and paint	131
	7.17	'Microporous' exterior stains and paints	133
	7.18	Non-film-forming finishes	133
	7.19	Exterior paints	134
	7.20	The durability of exterior finishes	135

7.21	The effects of lighter or darker colours	136
7.22	Chapter summary	137

8   Principal Softwoods Used in the UK   **139**

8.1	European redwood (*Pinus sylvestris*)	139
8.2	European whitewood (mostly *Picea abies*)	141
8.3	Sitka spruce (*Picea sitchensis*)	142
8.4	Western hemlock (*Tsuga heterophylla*)	142
8.5	'Douglas fir' (*Pseudotsuga menziesii*)	143
8.6	Larch (mainly *Larix decidua* and *L. kaempferi/L. leptolepis*)	145
8.7	'Western red cedar' (*Thuja plicata*)	145
8.8	Southern pine (*Pinus spp* – principally *Pinus elliottii* and *P. palustris*)	146
8.9	Yellow pine (*Pinus strobus*)	147
8.10	'Parana pine' (*Araucaria angustifolia*)	147
8.11	Species groups	148
8.12	Spruce-pine-fir	148
8.13	Hem-fir	149
8.14	Douglas fir-larch	149

9   A Selection of Hardwoods Used in the UK   **150**

9.1	Ash, American (*Fraxinus spp*)	151
9.2	Ash, European (*Fraxinus excelsior*)	152
9.3	Beech, European (*Fagus sylvatica*)	152
9.4	Birch, European (mainly *Betula pubescens*)	153
9.5	Cherry, American (*Prunus serotina*)	153
9.6	Chestnut, Sweet (*Castanea sativa*)	153
9.7	Ekki (*Lophira alata*)	154
9.8	Greenheart (*Ocotea rodiaei*)	154
9.9	Idigbo (*Terminalia ivorensis*)	155
9.10	Iroko (*Milicia excelsa*)	155
9.11	Keruing (*Dipterocarpus spp.*)	156
9.12	Mahogany, African (*Khaya ivorensis* and *K. anthotheca*)	156
9.13	Mahogany, American (*Swietenia macrophylla*)	156
9.14	Maple (*Acer saccharum*)	157
9.15	Meranti (*Shorea spp.*)	157
9.16	Oak, American red (principally *Quercus rubra* and *Q. falcata*)	158
9.17	Oak, American white (principally *Quercus alba*, *Q. prinus*, *Q. lyrata* and *Q. michauxii*)	158
9.18	Oak, European (mainly *Quercus robur*)	158
9.19	Obeche (*Triplochiton scleroxylon*)	159
9.20	Opepe (*Nauclea diderrichii*)	160
9.21	Sapele (*Entandrophragma cylindricum*)	160
9.22	Tatajuba (*Bagassa guianensis*)	161

9.23   Teak (*Tectona grandis*)   161
9.24   Utile (*Entandrophragma utile*)   161
9.25   Walnut, American (*Juglans nigra*)   162
9.26   Walnut, European (*Juglans regia*)   162
9.27   Whitewood, American, or Tulipwood
        (*Liriodendron tulipifera*)   163

10   Wood-based Sheet Materials   164
10.1   Plywood   164
10.2   The two fundamental properties of plywood   165
10.3   Basic types of plywood   166
10.4   Conifer plywoods   166
10.5   Temperate hardwood plywoods   168
10.6   Tropical hardwood plywoods   168
10.7   Plywood glue bond and 'WBP'   169
10.8   Exterior   170
10.9   Adhesives used in plywood   171
10.10   BS 1088 marine plywood   172
10.11   Plywood face quality   172
10.12   Appearance grading of face veneers   173
10.13   Conifer plywood appearance grades   173
10.14   Temperate hardwood plywood appearance grades   175
10.15   Tropical hardwood plywood appearance grades   175
10.16   Particleboards and wood chipboard   177
10.17   Flaxboard and bagasse board   178
10.18   OSB   178
10.19   Fibreboards   181
10.20   Hardboard, medium board and softboard   181
10.21   MDF   183
10.22   Chapter summary   185

Appendices   186
1   A Glossary of Wood and Timber Terms Used in the Timber
    and Construction Industries   186
2   A Select Bibliography of Some Useful Technical References
    About Wood   200
3   Some Helpful Technical, Advisory and Trade Bodies
    Concerned with Timber   201

Index   202

Preface

This is not really a 'text book' in the proper, academic sense of those words. Think of it more as a sort of helpful guide to wood and all its uses in construction. Of course, if you want to learn some more about the highly interesting and rather more technical subject of 'Wood Science' (which of course, is a perfectly laudable ambition), then there are many other far more academically-oriented books out there that you can get your teeth into. And if you look at the back of this book, you'll see I've given you a few helpful suggestions, along with the contact details of a number of technical, advisory or Trade bodies who can give you specific guidance on particular aspects of the uses of timber and wood-based products.

But, although this book is not meant to be highly erudite in the truly 'scientific sense', it is very definitely meant to be highly *informative*: so it will still tell you some legitimate and very useful facts about the best way to use wood, so as to get the best out of it. This book can, of course, be read by anyone with a positive interest in wood – be they an amateur or a hobbyist – but I have first and foremost aimed it the professional level: at the timber specifier or user. Nevertheless, it is my hope that even a serious amateur should get a lot out of it as well; because it is intended to try to put right a good many of the things that seem to go wrong with wood: and which are most often born out of a level of sheer ignorance about the material.

From my considerable experience, which has been gained throughout my very long and varied professional life, I have found that whenever things go wrong – such as in carpentry or joinery matters – it is *never* the fault of the wood. What has actually happened in reality is that someone, somewhere along the chain of specification, order, supply, installation and use, has done something that they really should not have done: and most likely, that was because they simply didn't know any better. And that ignorance about even the very basics of wood is the knowledge gap which this modest little book is designed to fill.

At the time of writing, I have been a consultant Wood Scientist and Timber Technologist for something over 35 years. And in that period, I have been involved with a great many highly interesting projects: all of them connected with the myriad applications of wood; and the majority of those concerned with its uses in construction. I have climbed up to the top of countless Cathedral and Church towers; I have crawled beneath the floors of numerous Nonconformist chapels (and a lot of other very old structures); and I have clambered amongst the roof timbers of many, many buildings, both historic and newly completed. Shakespeare's Globe Theatre and HMS Warrior were two of the more interesting projects needing some technical input from me; but amongst my hundreds of rather more mundane consultancy jobs, I have examined such things as kitchens in Kettering, decking in Doncaster, cladding in Cleckheaton and scaffold boards in Scunthorpe. I have inspected, selected, graded and rejected hundreds of thousands of boards, battens and planks of timber – both Softwoods and Hardwoods – in sawmills and woodyards as far afield as Montreal and Munich, Hull and Helsinki, Newcastle and New Zealand.

For many years, I was a Visiting Lecturer on Timber to both the Schools of Architecture and Engineering at the Universities of Newcastle and Durham. I have also delivered lectures in other parts of the world: such as on multi-storey timber frame construction at the University of Melbourne in Australia, and on wood preservation at the Centre for Advanced Wood Processing at the University of British Columbia in Vancouver. In fact in my time, I have carried out consultancy work on timber and wood-based panel products in over 30 different countries around the world. And I have also acted as an Expert Witness in hundreds of legal cases, large and small, where the only completely innocent party was the wood. The disputing parties were usually arguing the toss about poor specification, poor workmanship, or just plain ignorance of timber: any or all of which factors had then led to quite unnecessary losses in terms of money, time or materials.

Because Mankind uses wood the world over; but more especially, because the UK imports so many types of timber and wood products from every conceivable corner of the world, there is a huge amount of wood that is used in the UK, which never grew here. But it still needs care and thought in its use: and from the amount of time that I've spent teaching, investigating and explaining about wood to countless timber traders, builders, architects and engineers, I reckon to have a pretty good knowledge of wood in all its various guises. I understand intimately its cell structure and its properties; I know in detail about its types and species; and I am more than familiar with its grades and qualities. I understand almost intuitively how it works: and I am sure that I know how to make it work better, for everyone who wants to try and use it more thoughtfully.

But more than that: I have seen – far, far too many times – what people do wrongly with wood, which then makes this highly adaptable material perform badly, when it should work brilliantly.

As I hinted earlier, this book is intended to change all of the (often unintentional) bad behaviour on behalf of those who specify and use wood: by giving you, the reader, a basic – but I hope, very clear and workable – understanding of this unique and wonderful material. This book will arm you with the knowledge and information you need in a readily understandable form: thus helping you to get it right in any job you do with wood or wood-based products. Whether you are an architect, an engineer, a builder, a shopfitter, a timber salesman or simply an enthusiastic DIY-er: after you have read this book, I am confident that you will see wood in a whole new light, and that you will understand it a whole lot better than you did before. Furthermore, I am absolutely positive that, whatever your previous experience, you will learn something new from this book that will help you to use wood better and thus to get the best possible performance out of it. I hope you enjoy the journey you are about to embark upon. But most of all, I hope that in the future, you will use wood with care and understanding. Because it deserves it.

Oh: just before I leave you to get on with reading the book; let me give you a word or two of advice. The first couple of chapters – despite what I said right at the beginning of the introduction about the relative simplicity of this book's contents – might seem to be a bit too 'Wood Science-y' for your taste: but please try to resist any temptation to skip them. The information that I've included for you in those early stages is meant to establish some essential 'building blocks' of wood knowledge that you really ought to keep in your head, as an absolute minimum; and which will help you to better understand the more practical stuff that comes up in the later chapters.

So please give the earlier, harder bit a go: you might even surprise yourself as to how interesting it actually is! And if you feel that you really can't slog through all the technical details, then I suggest that you go straight to the Summary at the end of each chapter – so at least you can check up on the absolutely vital bits. But if you don't even want to do that, then you'll only have yourself to blame when the next court case goes against you, and you're then faced with a very large bill. Which will very likely happen, sooner or later.

Acknowledgements

I must just add a word or two of thanks to a few people who have helped me with the preparation of this book. I owe a very large debt to Gervais Sawyer, who helped tremendously with preparing many of the photographs. Also, I would like to thank my son, Neil Coulson, for converting all of the coloured pictures to high definition Black & White, as required by the publishers. My thanks also go to Iain Thew of TFT Woodexperts for preparing most of the Tables; and finally, I must thank all those who have contributed particular photographs, and who have also generously allowed me to use their copyright within this book: BSW Timber plc, Osmose Ltd, Canada Wood UK, Bob Caille, and Simon Cragg.

Jim Coulson
Bedale, North Yorkshire, March 2011

1 Wood as a Material

The very first thing to get absolutely clear at the start, is that there is no such thing as 'wood'! Of course, there is the stuff that grows on trees (or rather, the stuff that grows *inside* trees): but what I mean to say here, is that there is not one individual, unique and single substance that can simply be referred to just as 'wood'. There is no one, unique material that will do every single job without any problems and with no prior thought, no matter what you might require it to do for you.

The stuff that we know as 'wood' – and as most laymen are apt to use that term – is merely a catch-all word that covers a whole range of possibilities in terms of appearance and abilities. From the hard-wearing to the hardly worth bothering with: or from the very strong and durable to the very weak and rottable. So, in this book, I aim to show that any given species of wood is very different in its properties – and therefore in its usefulness – to some other vaguely similar sort of wood, but which happens to be of a different species.

An obvious comparison could be with the idea of what we mean by the word 'metal'. If you should go along to a stockist of metals, then the first thing you're likely to be asked is exactly what job you intend to do with that 'metal'. And the answer to that, in turn, will govern the likely properties that you will want that 'metal' to possess. Do you require it to have a high tensile strength, or a good degree of ductility, or a shiny surface, or something else? And if you don't specify more precisely what you need this particular 'metal' for, then you may be offered a whole range of possibilities: ranging from steel, to brass, to copper – or tin, or lead, or mercury (which is liquid at room temperature) or even calcium (yes, although it's in your bones, it's a metal!). All of these genuine 'metals' are very different from one another, with huge variations in their physical and chemical properties; but all of them fit that initial, vague and general description of being a sort of 'metal'. So why should the situation be any different when it comes to wood?

A good question to ask would be: 'Why do so many people assume that 'wood' is all that they need to specify'? Even those who take more care about what they do or write, often think that they've done enough by asking just for a 'hardwood' or a 'softwood' – as though that somehow defines more accurately

Wood in Construction: How to Avoid Costly Mistakes, First Edition. Jim Coulson.
© 2012 John Wiley & Sons, Ltd. Published 2012 by John Wiley & Sons, Ltd.

the properties that they require in their material. But even that apparent improvement in the material's description is simply not enough, as I hope this book will show.

Every single, individual species of wood has certain very specific properties and therefore, it must follow, certain potential uses. But it also has certain other things about it that we might do best to avoid, or at least restrict: and those individual properties of this immensely variable material will then be subtly – or perhaps greatly – different from one species of wood to another. In essence, no two 'woods' are the quite same as one another; just as no two 'metals' are quite the same. And quite often, the differences in performance between different wood species can be very large indeed.

Sometimes, of course, these differences in properties are quite minor; and they will not significantly affect the outcome, where one species has been used instead of another. But sometimes, the differences between alternative wood species can be absolutely vast – such that it would be the equivalent of using chalk instead of cheese. (I know nobody builds with cheese – but sometimes, they might just as well, for all the good it does!)

There are at least 60 000 (and still counting) different species of wood in the world, which have so far been discovered and described by botanists or by Wood Scientists: so you should now begin to see that you really do need to know a whole lot more than perhaps you thought you needed to, in order to begin to understand exactly what *sort* of 'wood' you should be asking for. And, of course, what you should really be using.

But it's not *only* a question of the wood species – vitally important though that is. The Quality and the Grade of the timber that are to be used are also very significant factors in getting the best performance from timber, at the best price: as are a number of different processes and treatments that can (and quite often *should*) be done to the timber, once its wood species and final quality have been decided upon.

Some of these other processes are: moisture content (drying), treatment (preservation), finishes (paints and stains) and taking care of the timber during delivery and storage. All of these things are, in my humble opinion, quite essential factors in getting a good job done properly, when using timber. Not to mention all the additional complexities that are involved in specifying and using wood-based board products, such as plywood or chipboard or MDF. I will explain the most important of these different factors and different processes in greater detail, in some of the later chapters. But for now, I want to begin the process of your timber education by looking at what wood is actually made of.

1.1 Cellulose

All wood cells are made predominantly from cellulose. It's true that both the chemistry and physics of wood are somewhat more complex than this simple statement would imply; but I don't need to go too deeply into the chemistry

$$CELLULOSE = C_6H_{10}O_5$$

Light acts upon water and Carbon dioxide thus:-

$$Light \Longrightarrow H_2O + CO_2.....^n$$

This is photosynthesis

Therefore $5 \times H_2O + 6 \times CO_2 = H_{10} + O_5 + C_6 + 12O \; (= 6 \times O_2)$

In other words, 5 × Water + 6 Carbon dioxide = 1 × Cellulose + 6× Oxygen molecules

Alternatively, 6 × Water + 6 × Carbon dioxide = $C_6H_{12}O_6$ (Sugar) + $6 \times O_2$

This changes in the growing tree to

$$C_6H_{10}O_5 + H_2O$$

(cellulose)   (water) – absorbed into the tree

Figure 1.1   How trees make wood and oxygen

and physics here, in order to get you to appreciate the wonderful properties of this unique material. For now, suffice it to say that the main ingredient of wood – and therefore what gives this natural material most of its significant properties – is the organic substance called cellulose.

Cellulose is made by (and within) the tree itself, using as building blocks the sugars and starches that have recently been manufactured in the tree's leaves: and these chemicals in turn were obtained by harnessing the energy of sunlight, under the influence of chlorophyll (that green stuff). In fact, every tree (and almost every living plant, for that matter) is a fantastic, natural chemical factory.

Simply by utilising nothing more complex than water, drawn up from the ground via the tree's root system, and then adding to it some Carbon Dioxide that is literally sucked out of the air, this wonderful 'chemical plant' then combines those most basic of ingredients, by simply shuffling the atoms and molecules around to make completely new ingredients out of them.

To make cellulose, the tree uses six molecules of H_2O (water) plus six molecules of CO_2 (carbon dioxide) to fabricate – as a first step – a single molecule of sugar ($C_6H_{12}O_6$). An extremely useful by-product of this chemistry – certainly so far as we humans are concerned – are 12 'spare' atoms of Oxygen (see Figure 1.1), which are helpfully released into our atmosphere in the form of six molecules of O_2.

After making itself a supply of carbohydrates (that is, sugar plus starch – which is really quite similar in its chemical construction: using as it does, only the atoms of H, O & C), the growing tree then uses this newly-produced food supply to manufacture cellulose ($C_6H_{10}O_5$) for itself: and as it does so, it then releases one 'spare' molecule of water. To complete the picture, this excess molecule of water is simply absorbed into the tree, so that nothing is wasted.

Having seen that the tree can conveniently make its own cellulose, we should then perhaps try to learn something about that particular substance. And the most fantastic thing about cellulose is that it is strong: very strong indeed. It is, in effect, a natural type of Carbon Fibre, invented by Mother Nature, long before Mankind ever got clever with chemistry.

It is the hugely strong chemical bond between the atoms of Carbon in the molecules of cellulose that gives wood its high strength. (These molecules are called, by chemists, 'long chain' molecules, because of their highly-organised, elongated and linked-together structure.)

Cellulose (and therefore wood) has, as I've just said, very high strength, which comes from the linked atoms of Carbon in its molecular chains. This amazing strength was shown way back in the 1960s: where an experiment was carried out at a major university, to prove just how incredibly strong wood can be. The experiment consisted of pulling apart two equal-weight strands: one made of European pine and one made of a high-tensile steel wire, using a special machine, called a 'tensometer' (which pulls things apart in tension). Then they measured the force that it took to snap each strand: and from this experiment, it was demonstrated that (weight for weight) wood is actually *stronger* in tension than steel!

However, the picture is not quite as straightforward as perhaps I've implied, when it comes to establishing exactly why and how wood is so strong. As well as knowing its chemistry: that is, that wood is made up of very strongly-linked molecules of cellulose, we also need to consider the *physical* structure of wood when we are looking at how it performs when we actually try to use it to do any job with. So I now need to tell you about the way wood is – quite literally – put together, in order that you can properly understand how best to use it.

1.2   Grain

Trees (and therefore of course, wood) have an inherent 'grain' structure. Grain is one of those common yet very over-used words, that laymen love to bandy about all the time when referring to wood in all sorts of ways: not least when describing its appearance (which is wrong). The word 'grain' has a very specific meaning: so it is important that I should help you to use this term correctly from now on.

First of all, what grain is *not* is that nice, wavy (or sometimes stripy or curly), and thus often highly decorative pattern which we so often see on the surface of a piece of planed or sawn timber. I wouldn't mind betting that most of you have used the word 'grain' in that context: and I suspect that perhaps many of you still do.

But that's not right. The correct name for this nice, decorative surface pattern on a piece of timber is the word 'figure' (see Figure 1.2). Figure can often

Figure 1.2   Example of figure (pattern) on the surface of timber

(although not always) show us what the *real* grain is up to; but it is decidedly *not* the same thing as the 'grain' of the wood. Sometimes, mis-reading the figure and thinking it is the grain can lead to physical damage: and sometimes it can lead to unnecessary rejection of the timber, for example when undertaking strength grading (a topic that I will discuss in a later chapter).

So, if it is not the pattern that you can see on the surface of the wood, then what exactly *is* grain?

Well, in my book (literally, as well as metaphorically!) the term 'grain' specifically relates to the direction of the wood fibres: that is, the way they grow up and along the trunk of the tree; or the way they are aligned along the length of a board or a plank of wood (see Figure 1.3). The principal vertical (or longitudinal) cells in the tree trunk – which for now, we'll refer to simply as 'fibres' – are relatively long (a few millimetres in softwoods) but they are very narrow, and they generally grow quite straight: along the main axis of the tree's trunk or stem.

These basic wood cells grow in the form of hollow tubes: which have a relatively thin cell wall, and with a hole (known as the cell 'cavity' or 'lumen') that runs all the way down their middle. In the living tree, this lumen or cell cavity is full of sap. But when a tree is cut down, the sap dries out (sooner or later), leaving the 'dry' wood essentially as a network of relatively long but narrow, hollow tubes, full of air. (I want to come back to the detail of the correct drying of timber later.)

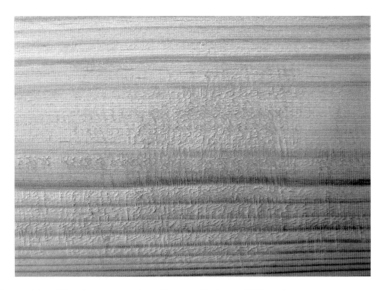

Figure 1.3 Wood surface showing grain (wood fibres)

These tube-like 'wood fibres' all point more or less in the same direction (along the tree trunk). So please remember from now on, that you should (and I definitely will!) only use the word 'grain' to mean one thing: 'the direction of the wood fibres'.

You should now see that, if we cut up a tree in a good and efficient way, such as in a sawmill, we will (hopefully) find that the wood fibres that were in the tree will line up so that they are more or less parallel with the long axis of the board or the plank of wood that we cut out of that tree. And if the cutting has been done well, then the grain will run pretty straight along the length of that piece of timber. I hope that you can now understand that those strong molecular 'chains' of cellulose (which, as I said earlier, make up the main substance of the wood-fibre walls) are then able to contribute their very high 'chemical-bond' strength to the physical 'along-the-grain' strength performance of the wood. So it is immediately possible to state one certainty about wood, and that is: 'the straighter the grain, the stronger the piece of timber'.

Unfortunately, though, there is always a plus and minus where wood is concerned (as you will see several times in the course of this book). The 'plus' is that the cellulose makes wood very strong when loaded along the grain. But the 'minus' with wood is that all of those long, thin wood cells that are in the tree or along the plank of timber, are stuck together, side-to-side, by a natural sort of glue, known as 'lignin'.

Lignin is a very complex chemical, whose structure is still not fully understood: but one thing that we *do* know about it, is that it is not very strong in tension. Therefore all of the wood fibres can be pulled apart relatively easily

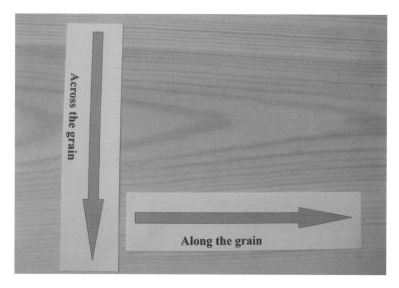

Figure 1.4   Directions, along and across the grain

in their side-to-side direction: and this sideways orientation is what we usually refer to as being 'across the grain'.

So, because of both its chemical make-up (the cellulose and the lignin) and its very particular physical structure (a whole bunch of long, very strong, tube-like 'fibres', which point along the line of the grain, but which are nonetheless stuck together relatively weakly) wood ends up being a very unusual material, in terms of its strength performance. We say that wood is *anisotropic* – which is a posh word, that basically means: 'it behaves differently under different directions of loading' – as I'll now explain.

Consider a brick, perhaps, or a block of concrete. Load either of them (by, for example, squashing them beneath the weight of a wall in a building) and they will resist that load – which is a compressive force – to the best of their ability. And, as you might reasonably expect, the brick or the concrete block is capable of resisting that load pretty equally, in all the three usual directions of width, breadth and depth. By the same token, a steel joist will be more or less equally strong when it is loaded in bending or in tension, in each of those three directions. In fact, just about all of our building materials, by and large, behave more or less equally in terms of their strength, in all of the different directions of loading. All of them, that is, except for wood: and that is because of wood's highly 'anisotropic' nature, which it gets on account of its special 'grain' structure.

Compared with our other building materials, wood really *is* very unusual. It is – as I've just explained – incredibly strong *along* the grain (and remember, that is along the direction of its fibres). But wood is very weak *across* the grain (that is, sideways, across both the width and the breadth of a timber member) (see Figure 1.4). This strength difference – taken as an average amongst most common wood species – is about 40 times greater in tension *along* the grain than it is *across* the grain. Now just think about that for a second: 40 times! That's an incredible difference in the behaviour of one single material: and one that is dependent only upon its direction of loading.

So that's why everyone who uses timber should always try to use it in a way in which they can capitalise on its long-grain strength; whilst at the same time, trying to minimise its cross-grain weakness. And they should do that by making sure that any imposed loads are carried as much as possible along the grain; and never, as far as possible, across the grain. However, strength alone is not the only property of wood that may be influenced by the direction of the grain.

1.3   Dimensional change in wood

Wood reacts with moisture: or to put it more accurately, it reacts to *changes* in its moisture content; which in turn are influenced very strongly by the relative humidity of the atmosphere that the wood may be used or stored in. I will talk about water in wood in a great deal more detail later; and I will explain the vital need to get things right, in Chapter 3. But for the moment, I want to touch on just one essential concept, that is related directly to grain orientation: and this is known as 'movement' (see Figure 1.5).

When wood loses or gains moisture from within its microscopic cell structure, the wood fibres will either come together a little bit, or shift apart a little bit, in their side-by-side relationship. (Don't worry at this stage *why* that happens: I'll explain it fully in Chapter 3. For now, concentrate on the fact that it really *does* happen.) Therefore, please take it as a fact that wood swells or shrinks, depending upon whether it respectively gains or loses moisture.

But this change of dimension (and that's essentially what it is) only happens to any appreciable extent *across* the grain. Wood does not swell or shrink to any significant extent *along* the grain (at least, not under normal circumstances of use and with 'normal', healthy wood). But neither does wood change dimension – in any direction – in response to any normal changes in temperature: so that is also a very useful property possessed by this unique material.

By now you should see that, as well as needing to know which direction of the grain is orientated, in order to use wood in its best (strongest) direction, you also need to know which way its grain is oriented, in order to make allowances for any swelling or shrinkage that may happen to timber components in service. So by understanding wood better, and by looking at it with a more experienced eye, if you can tell when you're dealing with timber in its 'long-grain' orientation, then you won't need to leave any expansion or 'movement' gaps in that direction. But if you find that it is being used in its 'cross grain' orientation, then you'll now know that you – or someone – must leave adequate movement gaps in that direction, to avoid later problems. (As I have said, the exact details of all this will be explained

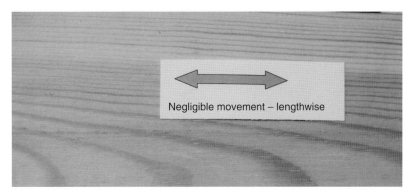

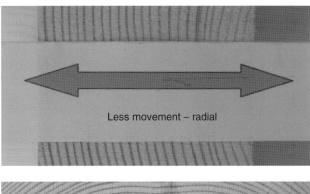

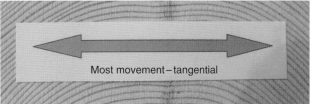

Figure 1.5   Movement

in Chapter 3 when we discuss Water in Wood: but at the moment, I am simply concentrating on the essential properties that are common to the grain of all wood, regardless of species.)

Before I leave the subject of grain – at least for the time being – there is another important property of wood related to grain orientation that I'd like to cover: and that is its ease or difficulty of machining.

Straight-grained timber (as you should now know, that is a piece of wood whose fibres are all nicely parallel with the long straight edge of the board or plank) will be much easier to machine than will be timber whose fibres 'stick up' out of the surface at an angle (see Figure 1.6). If they come up to the surface at a relatively steep angle (or sometimes even at several different angles!), then these fibres will more easily catch on the cutters of a planer, or on the teeth of a

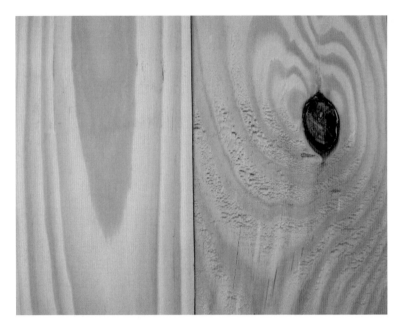

Figure 1.6   Two pieces of timber: one straight-grained, showing smooth surface and another with a grain problem, showing a 'splintery' surface

saw; and so they will make lots of splinters and give the timber a very rough finish. That is yet another reason to appreciate just exactly what the grain of wood is, and to find out where it's going and what it's doing, within any piece of timber that you are hoping to use.

But that's not all you should know: not by a long way. Apart from its grain structure, there are a good many other things to know about wood in general, which can help you to understand its unique properties. So let's now take a closer look at the cut end of a log.

1.4   Heartwood and sapwood

In many wood species (but be warned: not in all of them, by any means) you may see a distinct change in the colour of the wood tissue at the tree's centre – or heart. This central zone of wood tissue is known as the 'heartwood' and it forms the oldest part of the tree. That's because trees grow bigger by adding on layers of wood on the outside of the stem, just underneath the bark (that's the reason why trees have 'growth rings' in the first place). So therefore, the *newest* wood tissue is that wood which has recently formed just under the bark; whereas the *oldest* wood tissue is to be found right in the log's centre – dating back to when the tree was just a sapling. After a few years (and of course, exactly how long this 'few years' may be depends on various factors: the tree's particular species, the local climate where it is growing, its local forest habitat, and many other things), this central zone at the heart of the tree trunk simply shuts down; and after that, the heartwood takes no further part in the day-to-day growing life of the tree.

But the wood in the heart of the tree doesn't rot, or do anything strange (unless the tree is very, very old and thus over-mature: in which case the heart may eventually rot out). Normally, the heartwood simply closes itself off and takes no active part in conducting the sap up the tree; and it then leaves that job to the outermost few layers (that is, the most recent few years) of wood-cell growth.

So what about this outer zone or band of wood tissue? Well, this is the bit of the tree that still plays a full part in its day-to-day life: and it is known as the 'sapwood' (because it conducts the sap, of course!). Please note that the sapwood of any tree is generally pale in colour and in many wood species it is visibly distinct and separate from the heartwood in any tree. But – importantly – the sapwood in any tree has precisely the same basic cell structure as the heartwood: because it will one day become heartwood; when and if the tree carries on growing outwards and closes down more layers, as it gets bigger and bigger if allowed to keep on growing (see Figure 1.7).

The principal difference between sapwood and heartwood is that the heartwood has been deliberately shut down by the tree, in order to conserve its energy. But this one functional difference is nevertheless vitally important to us as wood users, because it can affect the properties (and therefore the potential behaviour) of each and every piece of timber. Every single piece of timber can potentially contain varying amounts of either sapwood or heartwood within it – and this is especially true of softwoods, whose structure will be described in Chapter 2.

Sapwood – by its very name and nature – contains (or in dried-out wood, it once contained) the 'sap' – which is the juice of the tree – and which carries both the tree's essential foodstuffs and its waste products. Sap is (or was, if we're talking about dried-out wood) very wet, and it is very rich in carbohydrates – those sugars and starches – which other living things really like to eat. Other living things, that is, like moulds, stains and rots. And sometimes, even other things like beetles and 'woodworm' (which, you'll not be surprised to hear, is not actually a worm).

Therefore you need to be aware of the fact that sapwood – under any adverse circumstances, such as very high levels of moisture – will be extremely prone to discolouration, or worse, if the high moisture level continues for some time: because then it is seriously at risk of being eaten by decay fungi. (Not to mention insects.)

Heartwood, on the other hand, because it has been closed down by the tree, contains much less moisture and (depending on the species) it also contains much less appetising stuff for those nasty organisms to eat. But here's the rub: and I hope you noticed that I used that simple but all-important phrase, 'depending on the species'. Different individual species of trees can manage to do quite amazingly different things when laying down their heartwood.

Figure 1.7   Log ends, showing heartwood and sapwood

Some species (such as the White Oaks, or American Mahogany, for example) manage to convert the residues that were left behind within the heartwood into quite different and more complex organic chemicals – such as the Tannins in Oak. Or they may accumulate deposits within the heartwood, made up from the chemicals and other things that the tree picks up out of the ground – such as the Silica deposits in Iroko. And yet, some wood species do nothing of the sort; and so they have no fancy chemicals stored within their heartwood to help them resist potential attack by bugs, beetles and rot.

The heartwood of certain particular wood species will contain what we Wood Scientists call 'extractives' (because they can be extracted from the timber by means of solvents, or steam, or some other process). And the presence of these extractives will very often change the colour of the heartwood: sometimes making it a shade or two darker, or even changing its colour completely.

But the really clever bit, as I hinted at above, is that those timbers which contain particular extractives within their heartwood end up being resistant to (or sometimes, virtually immune from) attack by decay fungi, and in some cases, even by wood-eating insects. And when a timber has such chemicals or deposits stored within its heartwood that help it to resist being attacked and eaten, we say that the timber has some level of 'Natural Durability' – and this is another of the fundamental properties of wood, which varies from species to species.

1.5   Natural durability

Please be careful with this term 'durable'. In the context of wood, it does not mean 'hard wearing' or 'strong': it *only* means that the timber *may* have a level of resistance to (usually) rot, which can then help it to last longer under adverse conditions, such as when exposed to high levels of moisture for long periods. And I make no apologies for labouring the next point, because I need you to be really clear about it.

The heartwood of certain wood species *may* have a degree of natural durability (decay resistance): but this will be entirely dependent upon the particular species of wood (and *not* whether it is a 'Hardwood' or a 'Softwood' either – as I'll discuss in Chapter 2). And this Natural Durability will not be directly related, in any obvious way, as to how dark the colour of the heartwood is. For example, the heartwood of Oak is a light, honey-coloured timber, yet it is rated as being in the second-best rank of rot resistance; whereas Scots Pine heartwood is red-brown, but it is only in the fourth-best rank of Natural Durability.

Remember that the sapwood, as I stated just a little while ago, has no great amount of natural durability whatsoever. None. Not in *any* species. And that can sometimes give us quite a headache: because of the very real possibility of there being a fundamental difference in the durability of the sapwood (i.e. hardly any) and the heartwood (perhaps very high) *within the same piece of timber*.

So you will need to consider very carefully the conditions under which you are intending to use any specific timber. And after having done that, you may then have to decide that it is best not to use any of its sapwood; and so you may have to specify the *heartwood only*, of this timber that has a good natural durability. And that will mean that an additional process is needed, to remove all of the sapwood from the finished product. ('What about Preservation Treatment?' I can hear you asking. And I will answer that, I promise you: but in a later chapter. Remember, I'm dealing only with the fundamental properties of different wood species, in this chapter.)

I'm soon going to leave the properties of wood behind (though only for the time being, I assure you, since they are going to feature in every decision you make about wood from now on!). But before I do, there is one other basic property of wood that I'd like to mention, since it also affects the use of timber when it comes to applying preservatives. And in order to do that, I first need to tell you about another very special type of wood cell that is found within all trees: and that cell is known as a 'ray'.

1.6   Rays

I said, some pages earlier, that the main wood cells in a tree (and which I have for the moment referred to simply as 'fibres') are lined up vertically along the tree trunk, and – hopefully – more or less straight along the length of a piece of wood. And whilst that is true for the fibres, all trees need to have other, more specialised cells, which can help them to move the foodstuffs and waste products about, in and out of the trunk, in a horizontal direction as well. That is, from the point just under the bark to all the way inside: right to the very edge of the heartwood, within the 'active' sapwood zone.

And these horizontal cells 'radiate' out into the curvature of the tree trunk, just like the spokes do in a cartwheel: from the centre, towards the outside of the tree. That is why they are given the name 'rays' (see Figure 1.8). (By the way, I hope you can see that the heartwood was, once upon a time, sapwood itself, at an earlier stage in life of the tree: and so these ray cells can be found right throughout the tree trunk, and not just in the present-day sapwood.)

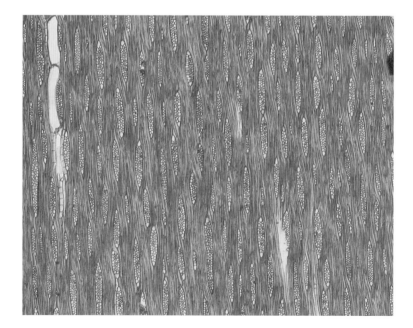

Figure 1.8   Rays, viewed end-on

As I have just said, these ray cells radiate out horizontally *across* the tree trunk, thus providing some very handy pathways in and out, in a sideways direction, *across* the grain. And now at this stage, I also ought to tell you the correct name for the two different versions of 'across the grain' – because they will become quite important in later chapters.

1.7   Radial and tangential directions

The direction across the tree trunk which goes the same way as the rays – that is, at right-angles to the growth rings – is called (for obvious reasons) the 'radial' direction.

But there is another way to travel across the grain; and that is by traversing a line which goes at right-angles to the radial: striking the growth rings across their curvature. Because this direction hits the circular rings at a tangent (remember your school geometry?) it is called the 'tangential' direction (see Figure 1.9). Please keep these definitions tucked away at the back of your mind: they will come in very useful later on, when we discuss moisture movement in greater detail.

But now back to what the rays help the tree to do. These ray cells, which radiate across the tree trunk, help to contribute to another property of wood, known as 'permeability'.

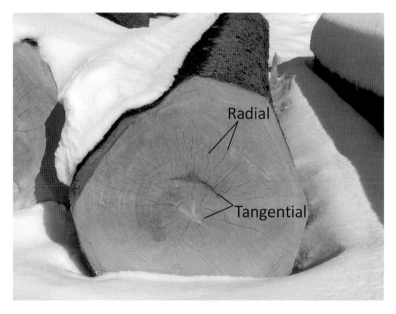

Figure 1.9   Log showing radial and tangential directions

1.8 Permeability in timber

The free movement of liquids in and out of the tree trunk is greatly helped by the rays. (There are also special openings – a bit like trapdoors – in the walls of the vertical cells; and these are called 'pits' and they are very important, especially in softwoods. But that's a bit too much like serious Wood Science for this book; so we'll leave things with just the rays, for now.)

Although we mostly use wood when the majority of the water from the sap has been removed (and I will cover Water in Wood in Chapter 3, as I said), there are occasions when we have a definite need to re-introduce a different type of liquid back into the wood cells. I am of course talking about Preservation Treatment. (I promised to tell you more about that later, also – and I will. But don't forget, we're still talking here about basic wood properties; and we're trying now to get to grips with the notion of permeability.)

So: imagine trying to pump some liquid back into the wood cells, without a series of helpful pathways to achieve it. Done that? Now imagine trying to pump liquid back into the wood cells, but this time with a whole series of wide-open pathways to help you to achieve it. Done that too?

Then you've successfully imagined the concept of permeability. All that this property involves is the ease or difficulty of penetrating the wood's surface and getting it to take up as much liquid (in this case a Wood Preservative) as we would like it to. And that is another basic property of wood.

But by now, it should come as no surprise to learn that some timbers are quite easy to treat, whereas others are frightfully difficult to penetrate. Some timbers are therefore classed as 'permeable' whilst others are classed as 'impermeable'. The varying levels of either the straightforwardness or difficulty in getting preservative treatment into wood, are subdivided into categories: ranging from 'easy to treat', to 'resistant to treatment' and right up to 'extremely resistant to treatment'.

As I've been saying to you all along: no two species of timber are the same, when it comes to their properties, or to using them: even to do exactly the same job with. Therefore, you really need to know what those differences between species are, if you're going to use wood successfully, time after time. So now, before I finally close this section on the Basics of Wood as a Material, I'd better just remind you what we've discovered about the properties of wood, so far.

1.9 Chapter summary

Wood is quite unlike most of the other materials that we use for building with. It is 'anisotropic' (it has unequal strength when loaded in different directions); and its principal strength lies along the *grain* – which I have defined as being the longitudinal direction of the main wood fibres. Wood gets this great strength (which, weight-for-weight, is stronger than steel) from the main constituent of its cell walls: which is a natural 'carbon fibre' called *cellulose*, and which has its extremely strong, long-chain molecules lined up along the main lengthwise axis of each long, thin wood cell.

This very special and unique grain structure of wood also affects its property of *movement*: which refers to the dimensional changes that occur in response to any changes in the wood's moisture content. Remember: wood only 'moves' (that is, changes its dimension) *across* the grain, and not *along* the grain: which is another unique thing about this material. (Steel and concrete, for example, expand by very large amounts, due to changes in temperature, and so structures made from these materials require expansion gaps every so often along the length of, say, a bridge or a railway track: whereas timber requires no longitudinal expansion joints at all.) And remember, too, that there are two different cross-grain directions: radial and tangential.

The nature of the wood's grain also influences how easily or otherwise we can work a length of timber. Straight-grained timber is much easier to finish well; whereas timber with wild or twisted grain will tend to machine more roughly and to splinter more easily; and thus it will require more time and effort to achieve a good finish. (You should realise too, that having a 'wild' grain will also result in the piece of timber distorting more, when it dries out.)

The next thing we learned about was that all trees have sapwood and heartwood: and it is the 'extractives' in the heartwood which sometimes (but not always!) help to give some timbers a greater resistance to rot and 'woodworm' attack. And we called this decay-resistant property its 'natural durability'. Another vital point to remember is that whilst the heartwood of certain species *may* have an increased level of durability, the sapwood of *all* species is *always* susceptible to such attack, whilst it is wet.

Finally, I told you about *rays* and the job they do within the tree trunk in a sideways direction. Since they form pathways in and out of the tree, *across* the trunk, they can either help or hinder us in impregnating timber with preservatives: and this property of any wood is what we call its 'permeability'. That's what makes different woods either easy to treat, or difficult to treat.

For the moment, that's enough about the basic properties of wood as a material. Now instead, it's time I explored the essential features of some of the different timbers that you're likely to use.

2 More on Timber: Some Stuff About the UK Timber Trade

2.1 'Wood' or 'timber'?

So far, I have been using the terms 'wood' and 'timber' with reasonable frequency; and you would be forgiven for thinking that these terms were more or less interchangeable. But if you go back and read carefully through what I've written so far, you should see that I have mostly used the word 'wood' to refer to the material that grows within the tree, and I have generally reserved the use of the word 'timber' to refer to a specific chunk of that material, such as a plank or a board. To put it another way: a floor joist, for example, is made of *timber* and a length of timber is made of *wood*.

Wood grows naturally, in billions of trees all over the world (and as a helpful by-product, it provides us with much-needed oxygen); and all of those trees are divided into literally tens of thousands of different species – as I've already explained in outline in Chapter 1. And yet only a surprisingly small number of those species that grow all over the world are actually used as different types of timber: that is, ones which are quite deliberately cut down as part of a forestry operation and then processed commercially in sawmills. (Most of the stories in the press about 'illegal logging' in various parts of the world, do not concern the Timber Trade at all: they are actually the result of land being cleared for other purposes, such as housing, or agriculture for local people. Yet it is the Timber Trade which tends to get a lot of unfair publicity from journalists and the media in general, who – quite rightly – object to the notion of trees being cleared away, willy-nilly. But the commercial Forest Industry – which effectively supplies all of the world's sawmills, and thus eventually our Timber Trade – does not 'sell off the family silver' in this way. Far from it. It treats its commercial forests as exactly that: something highly commercial and valuable, to be managed and kept going for the future. The act of planting four or five trees for every single one harvested, is quite normal in all of the well-managed forests in the Westernised world: and that has been the case for well over a hundred years.) I've now digressed somewhat … so let's get back to timber uses.

Wood in Construction: How to Avoid Costly Mistakes, First Edition. Jim Coulson.
© 2012 John Wiley & Sons, Ltd. Published 2012 by John Wiley & Sons, Ltd.

2.2   Wood species and timber trading

These few thousand (of the world's total number of wood species) which are used commercially, are often traded right around the world – quite literally: with timber from Sweden ending up in Morocco and timber from Canada ending up in Japan, for example. And all of these species are then used to make things: and the biggest user of timber in the UK if not the world, in volume terms, is the Construction Industry. This industry generally buys its wood from well-established Timber Importers or Timber Merchants (collectively called the 'Timber Trade', which I've been referring to).

Within the UK Timber Trade, there are a number of quite 'industry specific' terms that are used on a daily basis – and every industry has its jargon, as you probably know. So I'll start with the most basic terms and see if I can clarify them a bit more. (see also the glossary at the back of this book.)

2.3   Softwoods and hardwoods

One of the first things that anyone coming quite fresh to the use of timber will most likely hear people talking about, is one or other of the names used for two very basic sub-divisions that are used to describe all timbers. These are the (somewhat misleading) terms known as 'Softwood' and 'Hardwood'.

I'm sorry that I have to keep saying this sort of thing – since it looks like I'm nagging – and yet it really *is* important that you understand this properly. It is an uncomfortable and perhaps surprising fact, that these seemingly straightforward and apparently quite descriptive terms do *not* mean that any timber is either 'soft' or 'hard' to the touch. In fact, these terms really mean only one thing that can be regarded as entirely accurate: and that is, they will indicate *only* the basic sort of tree that any particular type of timber comes from. And *nothing* else.

Put simply, all of the so-called 'Softwoods' come from trees that we can describe very basically as conifers. These are, as the name implies, trees that usually bear cones: and they all will have needle-like leaves (see Figure 2.1). Furthermore, if we look closely (with a microscope) into the trunk of a conifer, we will see that its cell structure is much more primitive or 'basic' than the other type of tree (and which we'll get to in a moment). That is because the conifers – i.e. the Softwoods – evolved much earlier on our planet than did the other sort of trees. (Oh, by the way, if you're wondering why I said they 'usually' bear cones, one notable exception to this cone-bearing rule is Yew: which is indeed a Softwood and which has needle-like leaves: but which has small red berries instead of cones.)

In all of the conifers, the cones are always more or less open at their sides, so that the seeds within them are effectively unprotected, with no fleshy coat to cushion them or nourish them as they germinate. The usual phrase which describes this condition in plants is that they have 'naked seeds'.

Figure 2.1   Conifer (softwood) – showing needles and cones

All of the so-called 'Hardwoods' on the other hand, come from the only other basic type of tree that there is in the world: and that is the broadleaf. This type of tree – as the name clearly tells us – has 'proper' leaf-shaped leaves; and it also bears some sort of fruit (such as a cherry or a conker, for example), which both covers and protects the seeds within it (see Figure 2.2).

Now, if you want to be really posh and show off your plant knowledge to your friends, you can refer to the naked-seeded conifers as *Gymnosperms* and the covered-seeded broadleaf trees as *Angiosperms* – although if you don't want to seem too much of a know-all, it doesn't really matter: because you won't need to bother about that level of detail, in order to understand the essentials that I want to put across in this chapter.

The broadleaved trees that – according to the generally-accepted (although misleading) word that I gave you earlier – we call Hardwoods, are plants that evolved much, much later on our planet than the Softwoods did: about 150 million years later, in fact. So they had aeons of extra time at their disposal, in which to develop a much more complex sort of cell structure for themselves. Therefore the Hardwoods have ended up with many more different types of cells, which are arranged in many more different ways, than the rather more primitive Softwoods have managed to achieve.

That's basically the reason why there are relatively few species of Softwood in the world, and why they all have more or less similar characteristics to one another: because they had many fewer evolutionary characteristics to play about with. In contrast to that situation, there are thousands upon thousands

Figure 2.2   Broadleaf (hardwood) – showing leaves and fruit

of different species of Hardwoods, spread over most of (though not all of) the globe, and which manage to run the whole gamut of wood characteristics – from pale to dark, from light-weight to incredibly heavy, and from easily rottable to incredibly durable. And that is all because they had many more evolutionary ingredients to play with.

Another fact that is useful to know about Softwoods and Hardwoods, is that Softwoods in general prefer to inhabit the colder parts of the world; whereas Hardwoods are more comfortable in the slightly warmer (or often much hotter) places (see Figure 2.3). That's why you'll find that conifers are fantastically abundant in all of the more Northerly countries, such as Canada, Russia, Sweden or Finland; and they are also very abundant in most mountainous regions, such as in the Alps or the Rockies, where altitude tends to have the same cooling effect as a Northerly latitude does.

I'm generalising a bit here, since there are places in the Southern Hemisphere that also produce their own native conifers: but from a UK perspective, we don't really see the native Southern species of conifers in our timber yards. We do import Softwoods from the Southern Hemisphere, and yet with one notable exception (which is 'Parana Pine'), all of these Southern Softwoods are plantation-grown timbers that come from Northern Hemisphere species, and which have been transplanted into countries in the Southern half of the world. For the moment, I'd like to concentrate on our Northern timber trading partners, just to keep it simple.

Hardwoods, in contrast to Softwoods, are found both in the Temperate Zones (for example, in mainland Europe or in Southern Canada and the USA)

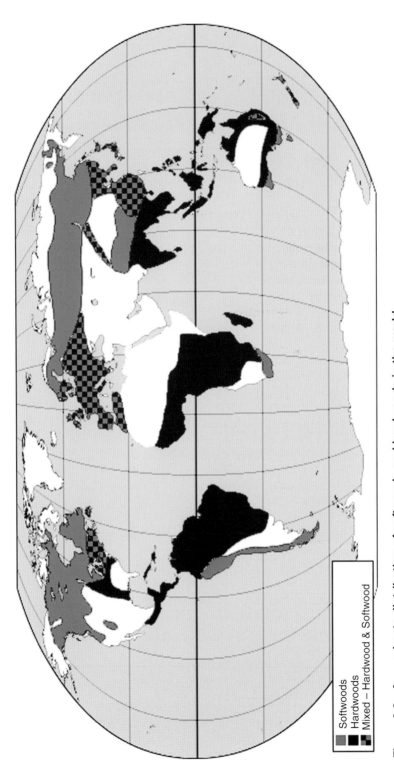

Figure 2.3 Approximate distribution of softwoods and hardwoods in the world

Softwoods
Hardwoods
Mixed – Hardwood & Softwood

and also abundantly in the Tropics: in other words, in places where it is either quite warm for a lot of the year or it is very, very hot for most of the time. The greater evolutionary extravagance of the broadleaf trees has also produced for us two distinct categories of trees: which we refer to (for obvious reasons) as either Temperate Hardwoods or Tropical Hardwoods. And each of these two groups tends to have some features within their timbers, that are common to all members of that group and which are not found as features in the other group … though it's not quite as simple as that (as I keep on saying!).

2.4 Some more detail on cell structure

I said, a little while ago, that the cell structure of Hardwoods is much more complex than that of Softwoods, which is true. This complexity also shows itself in the evolution of two further and quite distinct categories of Hardwoods; based not on their geographical location (that is, Temperate or Tropical, as I mentioned a minute or two ago) but on the very particular arrangement of their principal cell types. That means I now need to tell you in more detail, just what the wood cells in the two tree types are *really* called – and I will now stop using the catch-all term 'fibres' to describe everything in the tree trunk.

Softwoods, as I have said, are more primitive in their evolution and thus they are more 'basic' in their cell structure. This has resulted in their having only one main type of vertical cell, which then has to carry out both of the two very essential jobs, which the vertical wood cells are there to do for the tree. And these two essential tasks are: the conduction of liquid (i.e. sap) up and down the trunk; and the provision of structural support for the tree trunk itself.

You could think of a vertical wood cell in a Softwood as being a bit like a very, very small drinking straw: since it is essentially tubular and it has a thin wall, with a hole down the middle. It is this hole down the middle of the cell that undertakes most of the conduction of liquids up and down the vertical height of the tree; whilst it is the solid substance of the wall (and its relative thickness or thinness) that does the job of providing physical strength to the wood.

This very versatile and, so to speak, 'dual purpose' cell in a Softwood is more properly known as a 'tracheid'. So please forget the term 'fibres' in the context of Softwoods from now on, if you would (see Figure 2.4).

The other principal type of cell in Softwoods is the ray – which we've already encountered, in relation to its usefulness or otherwise in helping with permeability. In Softwoods, the rays are very, very small and quite narrow, so that you would need a microscope to see actually them.

Rays can also help the tree to store food; and so they are an example of another type of cell, which goes by the somewhat odd name of 'parenchyma'. (In reality these cells are rather soft and quite thin-walled; and they do not carry out any of the functions of the other specialised cell-types: that is, they are neither conducting tissue nor mechanical support tissue.)

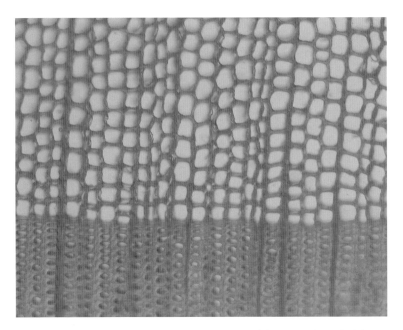

**Figure 2.4   Microscopic cross-sectional view of Softwood tracheids –
earlywood (upper) and latewood (lower)**

There are a few other very special types of cells present in Softwoods, such as resin canals and so on: but I'm trying here to keep it as simple as I can, by sticking to just the main cell types that will really have a strong influence on the essential properties – and thus the behaviour – of all the different wood species.

Well then, you might ask: what's so different about Hardwoods? Quite a lot, actually!

Hardwoods, as I told you earlier, are much more highly advanced in evolutionary terms: and for that reason, they have developed many more different types of cells with which to do all the different jobs that the tree requires of them.

In Hardwoods, the vertical conduction of liquid (sap) is done exclusively by a very specialised type of cell that is basically a very large diameter hollow tube – and on the scale of the microscopic wood cell, it is much more akin to a drainpipe than a drinking straw! This specialised vertical cell is found in all Hardwoods: and its sole function, as I have indicated, is to provide a passageway for liquids. It is properly known as a 'vessel' or a 'pore' (and we Timber Technologists tend to use the name 'pore' more often).

The other vertical cell in Hardwoods, which provides the structural support for the tree trunk, is really and genuinely called a 'fibre' (I've come clean about that term, at last!). So now you know that it is *only* the Hardwoods which have 'proper' fibres – although that word often stands in for wood cells in general, when people are talking about wood as a material and are not being very precise as to which 'proper' sort of wood cell they are dealing with.

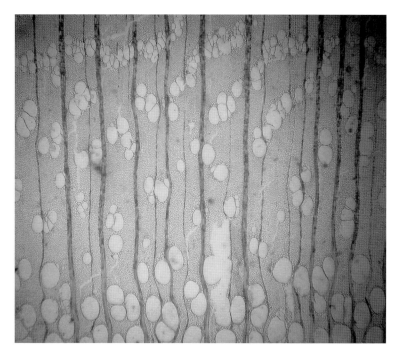

Figure 2.5   Magnified cross-sectional view of a Ring-porous hardwood, with wide band of pores forming the growth ring (bottom of picture)

I'm going to talk now about those *other* two sorts of Hardwoods – not the Temperate and Tropical ones, that are differentiated by where they grow in the world – but the two sorts that are described by referring to the way in which they have arranged their different wood cells.

One of the sorts of Hardwood has its growth rings clearly demarcated by a band (or a ring) of pores (those large, drainpipe-like cells) and this band of pores gives such timbers a very strong and clear growth-ring pattern on each of their cut surfaces. (This is a very good example of 'figure' – if you will recall the correct term for the pattern that shows up on the on the timber's surface.)

For this reason – that is, having a clear growth ring, that is strongly marked out by a band of large pores for each year of tree growth – such timbers are called 'ring-porous' Hardwoods (see Figure 2.5). To complete the picture, the fibres in this type of Hardwood tend to be grouped in bands too, albeit in another part of the growth ring. But more about growth rings a little later on.

The second (and in fact, the only other) type of Hardwood that is identifiable by what it does with its wood cells, has both its pores and its fibres more or less scattered – or 'diffused' as we tend to say – pretty well equally throughout the growth ring: so that there is generally a less obvious (and sometimes, not at all obvious) growth ring to be seen on the surfaces of the cut timber.

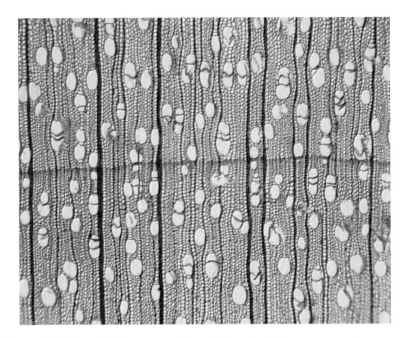

Figure 2.6 Magnified cross-sectional view of a Diffuse-porous hardwood with scattered pores and indistinct growth ring

Timbers which have their pores and fibres scattered all around are known as 'diffuse-porous' Hardwoods, because the pores are diffused throughout the entire cross-section of the timber (see Figure 2.6). This affects all sorts of properties of timber as well as just the appearance: the most obvious of which is texture.

Another difference between Hardwoods and Softwoods is that Hardwoods also have many more types of parenchyma (food storage) cells – and they are not just confined within the rays. These different sorts of parenchyma can be arranged in all sorts of ways (which, incidentally, can help us to identify many different timbers with a high degree of accuracy). This is yet another factor that helps us to prove that the Hardwoods are more highly evolved – and thus much more complex in their structure – than are the Softwoods.

Speaking personally however, I wish that those simple-sounding words 'Softwood' and 'Hardwood' had never become accepted as the norm: because all too often, they can mislead people into assuming that such timbers are either better or worse than they really are, when the names actually imply nothing of the sort. Softwoods are by no means all 'soft', or weak, or useless; and Hardwoods are definitely not all 'hard', or strong, or highly decorative. As I said before, the *only* thing that these two very general terms can tell you accurately, is the type of tree that the timber comes from: so if it is described as a Softwood, it's from a conifer; and if it's described as a Hardwood, then it's from a broadleaf tree. That's it: nothing else.

Those very basic terms cannot give you any other information about the properties of the timber that can help you to use it correctly. For such a level of detail, you need to know the individual wood species: and for that, you also need to know (at least in the first place) its 'scientific' name.

Now, although you might think that the 'scientific' names are a load of high-sounding twaddle, they are in fact extremely useful in helping you to decide whether or not a timber really *is* what you thought it was … and now, I think I'd better explain what I just meant by that last phrase!

2.5   Trade names and scientific names

The Timber Trade is of course a highly commercial industry: but it is one that is not at all 'scientific' (which comes as no great surprise) and so it has tended, over the years, to adopt a somewhat cavalier attitude to the names of the different timbers that it trades in. And many of these 'trade names' often bear very little resemblance to the reality of the world of genuine timber types and 'proper' names. For that reason, there has evolved a sort of 'two-tier' approach to the naming of timbers, whereby the Timber Trade works on a day-to-day basis with its 'trade names', but the world of Wood Science and Timber Technology works much more accurately, referring always to the 'scientific names'.

There is a British Standard, which was published some time ago, and which is there to assist the more conscientious specifier of timber to cut through all of the confusion surrounding 'correct' names. It is called BS 7359: 1991; and its full title is rather long-winded: 'Nomenclature of Commercial Species, including Sources of Supply', to be exact. But essentially, it is a list – quite a long list, in actual fact – of both Softwoods and Hardwoods which are used commercially in the UK. And it lists, very helpfully, both the timber's Trade Name and its Scientific name. And it also gives another, very useful list: that of 'alternative' names, which may have been used for the same timber in the past, or which might still be used for that timber in other parts of the world. So it is possible to discover that one individual species of timber may have three, or six, or ten possible trading names, depending upon who sells it and who buys it and where it came from in the first place.

There is also a European Standard, EN13556: 2003, which – in theory, at any rate – has now superseded BS 7359, at least in part. This European Standard also covers the names of timbers traded in Europe: but it has a number of significant omissions, as well as some rather perverse variants on Trade Names, which don't really reflect very accurately the real situation here in the UK. For that reason, I – along with a number of other Wood Scientists and Technical Timber Consultants in the UK – still prefer to use BS 7359, when it comes to deciding exactly *what* name a particular timber is best known by, here in the UK. (And, although BS 7359 is marked in BSI's Publications List as being 'withdrawn', it is still perfectly possible to get a copy of it from them.)

Anyway – I was talking about Trade Names and the tendency for there to be quite a few out in the wider timber world that may cause some confusion. Happily though, there is only one possible *scientific name* for any individual species of wood that we might wish to know about – which is why I always recommend that the scientific name should be referenced at least once in any specification, so as to avoid any hint of doubt or confusion over which actual timber or wood species is being referred to.

2.6   A cautionary tale about timber names

To be fair, most people don't actually need to constantly quote the scientific name on a day-to-day basis, when they are dealing with familiar woods that everyone uses all the while, in a familiar context. But there are times when an understanding of the difference between the name that the Timber Trade uses and what the timber really is, can be the difference between doing a proper job or making a very costly mistake. Take the fairly simple name 'Yellow Pine', as one particular example. And I'd like you to bear with me, in all the turns and twists of the next few paragraphs, because at the end of it all, there is a very salutary (and very expensive) lesson to be learned.

According to BS 7359 – and EN 13556, too – the UK Trade Name 'Yellow Pine' refers only to a particular species of true pine (we'll come back to the reason for that word 'true' in a little while); and this particular type of Pine is a wood species whose one-and-only scientific name is *Pinus strobus*. (Sorry, now it's time for another quick little digression, about exactly how Scientific names are 'put together', so to speak.)

Every living thing on this planet that has so far been discovered and fully described in 'scientific' terms, has been allocated into some form of 'family' network or system, that shows us what it is related to. This applies not just to trees, or plants; but to *everything* that lives, or has ever lived (yes, including dinosaurs). Bugs, beetles, animals, birds – you name it … or rather, some scientist somewhere has named it! And named it precisely and correctly.

In the world of trees, for example, all true Pines (that is, all of the different species of trees that share all of the usual characteristics that we associate with being a 'Pine') are put into the group called *Pinus*. (The proper name for this 'group' to which a species belongs, is called its 'Genus', by the way. And, since I'm being rather pedantic at the moment, I should also tell you that both the singular and plural of 'species' is 'species': its singular is *never* 'specie'!) But let's get back to naming trees and get on with the Pines.

There are other timbers in the world which the Timber Trade commonly likes to refer to under the name of 'Pine', but these wood species are definitely *not* Pines and they generally have very little in common with 'true' Pines – apart from a fancied or superficial resemblance, perhaps – because, as you will soon see, they have completely different scientific names. For example: the South American Softwood that is traded under the name of 'Parana Pine' has the

Figure 2.7   Southern pine (on LH) and yellow pine (on RH)

scientific name *Araucaria angustifolia*. So it should be immediately obvious that that name has nothing to do with any sort of real Pine, because it belongs in a completely different Genus – *Araucaria* – and so this means that 'Parana Pine is not a 'true' Pine at all. Thus when we use the timber from this type of Araucaria tree, we should not expect it to have any or all of the characteristics that we will find in a genuine Pine. I hope that's clear?

Now back to our cautionary tale involving the timber that is traded as Yellow Pine. By checking it out in BS 7359 or EN 13556, we can find out very quickly that this timber is a true Pine – simply from the use of the Genus *Pinus* as the first part of its Scientific name. But we can also tell *exactly* which individual type of Pine it is, from its one-and-only species name – '*strobus*' – and from that, we can then establish that it comes from Eastern Canada and that it is traded in the UK as 'Yellow Pine' or 'Quebec Yellow Pine'. (By the way, it's just as well that we can find out what the UK calls it over here, since in Canada itself, this same timber is known as 'Ontario White Pine'. Trust the Timber Trade to make things complicated!).

Unfortunately, the Timber Trade have made matters rather more complicated in another and somewhat more unhelpful way: by also trading in a timber from the Southern States of the USA, which should properly (according to BS 7359) be referred to as 'Southern Pine', but which the UK Timber Trade will insist on always calling 'Southern *Yellow* Pine'. This timber is indeed a true Pine; though it is not in fact just one single species of tree: and it certainly doesn't come from Canada (see Figure 2.7).

But that's another bothersome thing about the Timber Trade. They will often 'group' similar, related species together and sell them under one trading name – apparently to make their lives easier, without the need to separate them. And sometimes, they will even group quite dissimilar and completely unrelated species together. That sort of thing defies belief, if you're a Wood Scientist: but it's true – as we'll see when we discuss timber Grades, in a later chapter.

Still with me? I hope so. Now, the timber that is traded as 'Southern Pine' actually consists of more than *ten* different species of true Pine, which grow in the same forest areas in the Southern States. (This could be the reason why EN 13556 doesn't actually give 'Southern Pine' as a timber name, because it's a Group. And that's also one of the reasons why I prefer not to use this European Standard very much.) Anyway, one of the main members of the Southern Pine group is a true Pine called *Pinus elliottii*: also known locally as 'Slash Pine' (and this is what EN 13556 also calls it, but let's not go there just now). However – and this is the important bit of my story – absolutely *none* of the wood species included under the term Southern Pine is *Pinus strobus* (because that's the Canadian one, remember?).

'So far, so scientific,' I can hear you say, whilst probably trying to stifle a yawn: 'but what's the point of all this complicated naming stuff?' Well, if you'll stay with me a little bit longer, I'll enlighten you, I promise. And it's all to do with the properties and uses of these two very different timbers, and the confusion which genuinely arose from someone using those unfortunately similar (but quite different and separate) Trade Names.

Quebec Yellow Pine (which is the 'proper' Yellow Pine from BS 7359, if you recall) is a very mild textured timber, which is very easy to machine to a high standard of finish. It has virtually no resin in it, and it also has a very low index of 'movement' (that is, its dimensional change in response to moisture). For all of these reasons, it makes a very suitable timber for joinery: because it machines well, it doesn't easily react to moisture changes, and it has no resin which could ooze out and thus spoil paintwork or varnish, by making a nice surface rather sticky and unsightly.

Southern Pine, on the other hand, has virtually none of the above 'nice' joinery properties. It is somewhat coarser in texture and it has a very pronounced growth-ring figure. It is highly resinous and has occasional large resin pockets, which have a tendency to leak copious quantities of resin when the timber gets warm. It is also classified as 'medium movement' – which means it will respond much more to any changes in the relative humidity of its surroundings. So it is clearly *not* an ideal or first-choice joinery timber: in fact, it is really only recommended for construction, since it is pretty strong, and it benefits from a slightly enhanced natural durability. (We're nearly there now.)

Only last year (in 2010) I was asked to examine some timber samples which had been taken from a number of very expensive, craftsman-made doors, which had been specially manufactured for a cottage on the Isle of Skye. I concluded, from all of the identifying features I saw, that the timber which had been used was Southern Pine (and it was probably mainly *Pinus elliottii*). And now here is the point of all of this seemingly long-winded tale: the timber that had been requested by the householder was actually Yellow Pine (which of course is *Pinus strobus*). And sure enough, the joiner had gone to his usual timber supplier and had asked for 'Yellow Pine', but he had then been supplied with 'Southern Yellow Pine' by the Timber Trading company, who thought they knew exactly which timber had been ordered. But they didn't, of course.

The outcome of this cautionary tale is that these expensive doors – nicely made, and all that – have now shrunk far more than they should; they have cracked, and they are leaking resin from large resin pockets in several places. And so – unsurprisingly – the householder has refused to pay for them, and a very costly Court Case has been the inevitable result. And all because of that extra word 'Yellow' which the UK Timber Trade insists on inserting into the 'proper' name of Southern Pine. Yet all of the above problems – and all of the Court costs and wasted expense, too – could so easily have been avoided, if the specification had included the entirely unambiguous Scientific name *Pinus strobus* after the initial description of 'Yellow Pine'. Then the timber supplier would (or at any rate, should) have stopped to ask himself the question: 'What timber are they talking about here? I'm not sure on this one, so I'd better check on it.' And then (hopefully!) he would have discovered that it was 'Quebec Yellow Pine' and not 'Southern Yellow Pine' that was wanted.

If I may now take you back a few pages and remind you where we started with all this. I was saying that most people don't actually need to constantly quote the Scientific name on a day-to-day basis, when they are dealing routinely with familiar woods that everyone uses all the while, in a familiar context. And of course, that is largely true. But there are times when an understanding of the difference between the name that the Timber Trade uses and what the timber genuinely is, can be the difference between doing a proper job or making an almighty and very expensive mistake. Case proven, I think! So now let's move on.

2.7   Growth rings

I have referred to 'growth rings' quite a bit so far – and I'm sure you think you know what's meant by those words. And in a basic sense, you probably do. But I want now to elaborate on the whole idea of tree-rings and then explain how they can affect the quality – and can therefore influence the usefulness – of any particular timber.

Most people know that a tree puts on a ring of growth every year; and that if you cut down a tree and count its rings, you can immediately and accurately tell how old it is. Wrong! (Well, to be more precise: it's partly wrong and partly right … although it all depends. Let me explain.)

Temperate timbers: that is, wood from trees which grow in the areas outside of the tropics (and it can be either Softwoods or Hardwoods in this context) certainly *do* have 'annual rings' which relate directly to the years of the tree's life. And the annual rings in any Temperate Timber can be counted and so used to verify the age of any individual tree. So that bit of 'general wisdom' is correct, at any rate.

But tropical timbers – and this also includes any Softwoods which grow in tropical zones, as well as the more obvious Tropical Hardwoods – do not have 'annual rings' of growth. It is true that they *may* show some kind of 'growth rings', which are quite often indistinct; but these rings do *not* relate in any sense to the age of a Tropical tree. 'So what causes this difference?' you might ask: and the answer, more or less, is 'Winter.'

Outside of the tropics, the rest of the world has some sort of winter. Perhaps it's not always very harsh, but it is a sort of winter nonetheless; and it is a time of year when tree growth stops altogether. And it's this complete halt to a tree's growth that gives us the *annual* growth ring: a part of the year when no wood tissue is laid down at all, perhaps for quite some period of time. That period may be just a few weeks, as in the Western parts of the British Isles, where the Gulf Stream provides us with a very mild climate: or it may be more like six or seven months – or perhaps even longer – as it can be in the Northern extremes of Canada or Scandinavia quite close to the Arctic Circle: where both temperature and day length have a strong influence on growth – or lack of it.

This variation in the total length of the growing season, and therefore also in the proportionate tree-growing and wood-cell-producing time, has a marked effect on the nature and character of the growth rings that are produced. And also, of course, upon the wood that we get from those trees. In fact, a 'growth ring' in a Softwood has two quite distinct parts to it, which closely reflect the type of climate that it developed in.

2.8 Earlywood and latewood

In Softwoods (that is, conifers with a much more simple cell structure, remember) the growth ring consists of two separate but connected bands of tracheids (vertical softwood cells). In the early part of the year, when the tree has started up its growth again and thus requires a lot of sap to get up to the leaves (needles) very quickly, the new tracheids which the tree forms are very open, with a very thin cell wall and a correspondingly large lumen (that's the hole down the centre). But, if the tracheids were made just exactly like that for the whole of the year, then the tree would effectively be full of a lot of holes, with a very small amount of 'solid' wood substance around them, to keep the tree upright and so help it to resist the forces of the winter gales that will then try to blow it down. So the tree then does a very clever thing: it adapts its tracheids for strength.

At some point during that year – and this varies from location to location and from species to species (although geographical location is the more dominant factor) – the tree decides that it has to start thickening up its tracheid walls and also start reducing the hole down the middle of the cells, in order to give itself some good, strong structural tissue, in advance of the coming winter. So in this way, without ever changing the basic type of cells that it produces (i.e. vertical tracheids), a conifer can create two distinct bands of tissue within the same year of growth.

One band is rather soft and squishy, with no great strength, but it has the ability to transfer liquids in great quantities very quickly: and the other band is rather hard and strong, but with almost no ability to move liquids up or down the tree. And these two different bands, formed within the same year's growth ring, are are known as 'earlywood' and 'latewood' respectively – for obvious reasons, since they are related to the particular stage in the growing season when they were laid down within the trunk.

You will recall that I said a moment or two ago, that the particular amount of cell production varies, depending upon the tree's geographical location. Softwoods which grow in much colder regions, where the growing season is relatively short, will tend to form narrow growth rings that have a more equal proportion of earlywood to latewood – say about 50–50 or 60–40. But Softwoods which grow in very mild climates, where the growing season is very much longer, will tend to lay down a very wide band of earlywood; and then they will quickly finish off the year with just a short burst of latewood: thus producing growth rings that have a very high proportion of earlywood – perhaps as much as 90 % – with only about 10 % of latewood.

2.9 Rate of growth in softwoods

The ratio of earlywood to latewood – which is directly related to season length and therefore to climate – very much influences the characteristics of any Softwood timber that grows in any particular region. As we have just said, the colder the climate, the slower the tree growth; and thus the closer together are the growth rings (see Figure 2.8). Therefore the better the texture and the density (and also the strength) of the wood tissue: which means, in turn, a better overall quality in the log and a better yield of higher quality timber. But the milder the climate, the faster the tree growth: and thus the rings will grow much wider apart; and therefore they will give a much coarser texture and also a much lower the density (and also strength) to the wood tissue. This means, in turn, a poorer overall quality in the log and a much lower yield of high quality timber.

Putting it simply (from our UK perspective, at least), the more Northerly the growth of the conifers, the better the quality of the Softwoods they produce. And don't forget the influence of mountains and altitude, as I have said before: so Alpine Softwood timber (for example) with its higher altitude and colder climate – nearer to the snow line – will be of slower growth and thus of better quality, than will timber from conifers grown in the warmer lowland plains of Europe: which will generally have wider-spaced rings and a coarser texture.

We Timber Technologists refer to all this business of climate influence and ring-width – of fast and slow growth – as being the timber's 'rate of growth'. And I will discuss its effects in more detail later, in the chapter on Grades and Qualities.

But why, you might ask at this point, have I been at such pains to describe what happens within a growth ring; but only in relation to Softwoods? Isn't it

Figure 2.8   Fast-grown softwood (above) compared with slow-grown softwood (below)

just the same with Hardwoods? Well no, it isn't, I'm afraid! (I said at the start of this book that wood is very interesting: not least because it is not just one uniform material.)

2.10   Rate of growth in hardwoods

It's time I reminded you that there are two quite different types of Hardwoods, which are based on their cell arrangements: ring porous and diffuse porous, as I said earlier. And each of these two types of Hardwood looks and behaves quite differently from the other – and very differently again from Softwoods. So I'll consider the ring porous type of Hardwoods next.

I've just told you that a fast-grown Softwood will usually have low density and low strength, plus a coarser texture; whereas a slow-grown Softwood will tend to have higher density and better strength, plus a finer, more even texture. And that is true. However, the complete opposite is true when it comes to ring porous Hardwoods. And here's why.

Remember that the key feature with ring porous Hardwoods is that they have a band of very large pores, which clearly demarcates the annual growth ring. You can understand, I hope, that this is their version of earlywood (the part of the ring that conducts the sap). The band of dense fibres (with a few small pores scattered amongst them), which makes up the remainder of the

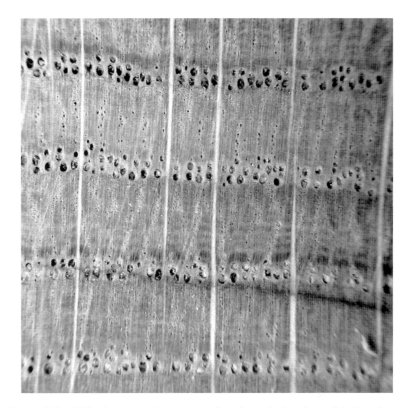

Figure 2.9   A Fast-grown ring-porous hardwood showing wide bands of latewood

growth ring, is their version of latewood (the part of the ring that gives most of the structural strength). And these two bands of growth – formed one after the other, in the same year – are directly comparable to the low-density earlywood and the more solid latewood that we find in the Softwoods.

But here's the key difference. In a slow-grown ring porous Hardwood, the earlywood band of large pores dominates the very narrow growth ring, thus leaving almost no space to fit in the denser band of fibres that should form the latewood: and so the timber ends up pretty much only full of holes, with no great amount of fibres for support. Therefore the timber will have a very low density, with all of the not-so-good things which generally go along with that situation.

In the exactly opposite way, a fast-grown ring porous Hardwood will have very wide growth rings, which then will have plenty of room for both the earlywood pores and the latewood fibres to fit into: so that the timber actually ends up being *more solid* – and thus a better quality of timber – just because it contains that many more fibres within its annual growth ring (see Figure 2.9).

So it is a curious – but nonetheless true – phenomenon, that fast growth is bad for producing higher strength and better quality in the Softwoods (conifers) but it is good for producing higher strength and more useful quality timber in certain Hardwoods (broadleaves): at least, so long as they are broadleaf trees of the ring porous variety. Examples of these latter timbers are: all of the true Oaks, Elms, Chestnuts and Ashes from any Temperate country, plus Hickory from the USA.

But of course, that's not all. Remember that there is still another type of Hardwood to consider: the sort whose cells are more evenly distributed throughout their cross-section – the diffuse porous Hardwoods.

These Hardwoods – which are by far the most numerous type of tree in the world – occur in both Temperate and Tropical regions. (But please don't forget that it's only the Temperate timbers – Softwoods and Hardwoods alike – that will show us genuine 'annual' rings; whilst Tropical timbers may or may not have very distinct growth rings; and these will not relate at all to a specific age of the tree.)

Given all of the above, you would be forgiven for asking: 'Do Temperate diffuse porous Hardwoods show us 'annual' growth rings, then?' And the answer is, yes they do. But their rings are not made up from a band of large,

Figure 2.10   Annual rings in a diffuse-porous hardwood formed by an absence of pores

earlywood pores: because these timbers don't have any 'earlywood'as such. (That only happens in ring porous timbers like Oak, remember. Stay with me on this!)

The annual rings that we find in diffuse porous timbers – such as Sycamore, Beech, Birch, Maple, Horse Chestnut and many, many others – are formed instead by a narrow band of some other type of wood tissue, whose composition varies, depending on the wood species. Typically it may often be a band of parenchyma (that is, those specialised food-storage cells), or perhaps it may be some coloured tissue deposits: or maybe just an absence of pores in that area (in other words, the tree forms a growth ring by just producing fibres alone, across a very narrow zone, at the start or end of the growing season) (see Figure 2.10).

Because of its very different formation, the rate of growth of diffuse porous Hardwoods is not subject to the same vagaries of density fluctuation that we saw with either the Softwoods or the ring porous Hardwoods; since the pores and fibres are pretty evenly spread throughout the timber's cross-section. Thus growth rates and seasonal or climatic variations have much less influence on the density and strength of diffuse porous Hardwoods.

But please, please remember this very important fact: just about *all* Tropical timbers are of the diffuse porous type – and they will not show you any clear growth rings and they do not have annual (yearly) rings. There was rather a lot of detailed stuff to think about there: so now I'd better give you a summary of the main points which I'd like you to concentrate on.

2.11   Chapter summary

First of all in this chapter, you learned that the terms 'Softwood' and 'Hardwood' don't tell you anything helpful about the properties of any timber that you might want to use – they only tell you the type of tree that they come from: those being either conifers or broadleaves. So you need to look beyond those terms if you want to know about individual wood species and to do that, you will need to find out the 'proper' name of the particular timber you are dealing with.

You then learned that the Timber Trade loves to confuse matters, by mixing up common timber names, regardless of which genus or species they might actually be in; and by giving essentially 'false' names to quite a lot of more common timbers, such as 'Pine' or 'Fir': and by introducing descriptions, such as colours, like 'red' or 'white' or 'yellow', when those terms are not always very accurate. So please watch out, when it comes to timber Trade names: and if in doubt, refer to BS 7359 (or, if you must, to EN 13556).

Finally, I told you about growth rings: how they are formed, and what effect the rate of growth can have on the quality of any timber grown under different climatic conditions. I showed you how Softwoods and Hardwoods differ quite fundamentally in the way in which they produce their growth rings: and how even the two basic types of Hardwoods have two completely different mechanisms for doing this. And as a very final point: please remember that you can't tell the age of a Tropical tree by cutting it down and counting its rings.

Now I need to look at Water in Wood in lot more depth (if you'll pardon the expression!).

3 Water in Wood: Moisture Content and the Drying of Timber

I have already touched on the fact that the uses of timber and wood-based products can be very strongly influenced – for good or bad – by their moisture content: that is, by the amount of water (in whatever form) they might contain under any particular set of circumstances. So now I'd better explain just exactly what is meant by 'moisture content' (which is generally abbreviated to 'mc').

Perhaps you've come across moisture content before? It is normally given as a percentage value: for example, 20% mc is the usual level for so-called 'Shipping Dry' in structural timber; or 15% mc may be quite usual for softwood flooring at the time of its installation. But 20% or 15% of what? What exactly does that percentage figure relate to? The *volume* of the piece of timber? Or its *weight*? And does 20% mc or 15% mc mean that the timber is wet or dry? And is it OK to use wood for all sorts of uses, at those sorts of mc levels?

In other words, the specifications which state moisture contents frequently use numbers that are quite often meaningless to the uninitiated. Or often, specifications will fail to give any meaningful numbers at all: and instead, they may use that dangerous and rather weasel-worded phrase 'kiln dried' (and which I will elaborate upon later). For now, I think that a bit of clarification is needed on what exactly is meant by the whole concept of moisture content.

3.1 The definition of moisture content

Moisture Content in wood is expressed as a percentage *by weight* of the water within the wood cells, compared with the *oven dry* weight of the wood cells themselves. (So it *is* weight, and not volume, of H_2O that determines its mc percentage.)

But what is this 'oven dry' weight that is used as the comparator? Think of it like this: take a chunk of totally dry timber (at zero percent mc, therefore) that now weighs exactly 1 kilogram – so that all of its weight is *only made up of* wood substance. But now, if it were to absorb 100 grammes of water, its overall weight would be 1.1 kg – agreed? Good: so we can now say that this chunk of wood has

Wood in Construction: How to Avoid Costly Mistakes, First Edition. Jim Coulson.
© 2012 John Wiley & Sons, Ltd. Published 2012 by John Wiley & Sons, Ltd.

a mc of 10% mc: because it has 10% extra weight (from the water) over and above its own perfectly 'dry' weight of the wood alone. The very simple formula for working out moisture content in any piece of timber is as follows:

$$\frac{\text{Wet Weight} - \text{Dry Weight}}{\text{Dry Weight}} \times 100 = \% \, \text{mc}$$

And anyone can quickly work out the mc percentage of our example chunk of wood above, by doing the simple arithmetic of this formula:

$$\frac{1.1\,\text{kg} - 1.0\,\text{kg}\,(\text{which gives } 0.1\,\text{kg})}{1.0\,\text{kg}} \times 100 = 10\%$$

The term 'oven dry' that I used above, is literally just that. To provide an accurate check on the moisture content (say when testing a load in a kiln), a small sample of the timber – usually about the size of a matchbox – is cut out of the middle of a much larger piece, taken from a stack somewhere, that is expected to be representative of the whole kiln load. (The reason why the middle of the sample piece is used is to avoid any influence from the ends of the piece: because timber always dries more rapidly from its ends, due to the nature of the wood-cell 'tubes', being cut at their open ends.) This small, trial sample is first of all accurately weighed; and then it is placed in a heated chamber (a small oven) at a temperature just above $100°\,\text{C}$ – to evaporate off all of the water – and it is repeatedly taken out and weighed at intervals, until it ceases to lose any more weight: thus showing that all of its water has now gone. This final figure is then its 'oven dry' weight. And the formula I gave you just now then is used to calculate what its actual mc must have been, just before it was dried it out. Simple, but effective – and surprisingly accurate!

3.2   Moisture meters

Of course, in everyday life, it would be entirely impracticable to cut samples out of every bit of timber that you needed to know the moisture content of; and then to dry and weigh them. So there needs to be a quick, reliable method to check on moisture content, in all sorts of places. And of course, there is one: it's called a moisture meter. But the first thing you should know about a moisture meter – especially if you're going to use it at all accurately and not let it confuse or mislead you – is that it *does not measure moisture!*

No, I've not gone mad – honestly! Although everyone always calls them 'moisture meters', they should really be called 'moisture content indication devices' or some such term: since in reality, they only measure one particular electrical property of wood – usually its resistance. And from that, a meter can give a reasonable *indication* – but only an indication – of the timber's present moisture content: and then, only at the precise spot where the meter was used.

And – because it is measuring electrical current flow and not actually water – a moisture meter can occasionally go wrong, or it can give false readings, or it can often be misinterpreted: or all three. As I will now clarify.

The first thing to note is that water conducts electricity – and that's the reason why, in the UK at least, they won't let us put light switches or electrical sockets in bathrooms (although it is strange that the Authorities will allow builders to do just that on the Continent. But that's another story: and it's much more related to our obsession with Health and Safety Laws, than it is to the Laws of Physics). But to continue: moisture meters capitalise on the fact that electricity and water have a relationship and, as I said earlier, that relationship is usually one of electrical resistance, which is measured in Ohms – or rather, in tens of thousands of Ohms, when it comes to 'dry' wood.

When a piece of wood is very wet, it has a very low electrical resistance; but when it is very dry, it has an extremely high resistance: so it will therefore not allow much current to flow through it at all (Strictly speaking, the current flows through the water that the wood contains, of course.) I will explain exactly what I mean by the terms 'wet' and 'dry' a little bit later; but for now, I want to concentrate on the that ways moisture meters can help you: but also possibly confuse or mislead you.

If you remember the crucial fact that the meter's probes (those are the metal points, that are either attached to the meter, or which you'll find on the end of a roving lead) are *only* measuring electrical current flow – and of course, the resistance to that flow – then you should not fall into the traps that can confuse the less well-informed meter users, who happily trust the meter readings without understanding their limitations. (The meter's, that is, not the person's – although either interpretation may be correct, when I stop and think about it!)

The standard, very short pins that are always supplied with every meter, are uninsulated: that is, they have bare metal shanks. And because of this, the meter cannot tell the you difference between the resistance of dry wood – wherever it may be located at any depth in a piece of timber – and wet wood, where the moisture may be situated only at the surface, or only at the depth, of that same piece. In other words, the meter will show you the same (high) reading when the wood's surface happens to be wetter than its interior, however deeply the pins are pushed into it, *even when the rest of the piece is acceptably dry.* That's because the wetter surface is being 'read' by the uninsulated shanks of the pins, which are directly touching it – thus creating a high reading – and this false reading then overrides any drier moisture level that may be present, a little deeper inside the wood.

The meter's standard pins are also, as I have just said, quite short – typically only about 5 or 6 mm in length at the most – so they are just not able to penetrate very far into the depth of thicker pieces of timber. This means that they cannot possibly show an accurate reading of the moisture content nearer to the core of the piece. So they cannot, therefore, indicate whether or not the wood is properly or fully dry; because they cannot tell if it still has a wet core. And unfortunately for the specifier and user of wood, both wetter cores and

Figure 3.1   Electrical resistance-type moisture meter with hammer probe

drier surfaces happen quite often – both during the initial drying process (that is, before the wood has fully reached its intended and scheduled final moisture content level) and under certain conditions of storage, delivery or use.

So how can you best avoid those false or inadequate readings, when using the short, uninsulated pins? The answer is really quite obvious: don't use them! Use instead, a 'hammer probe' (sometimes also referred to as a 'hammer electrode') which has much longer pins – generally about 25 mm or so in length – that are insulated along almost the whole length of their shanks, with only their very tips exposed (see Figure 3.1).

With these long, insulated pins, you can take readings of the timber's interior mc, without its exterior surfaces creating falsely high readings: such as when properly dried wood is exposed to a rain shower on the way to site. (It is quite common for the Clerk of Works or Site Agent to reject the timber delivery because he thinks it's wet through!) Or you can take a series of readings, at ever-increasing depths, further and further into the centre of the timber's cross-section (for timber thicknesses up to about 50 mm, at least), which will then build up a so-called 'moisture gradient' or profile of the mc at different stages throughout the timber's thickness. This will then show you whether the timber is getting increasingly dry, or increasingly wet, as you go towards its core. And that sort of data is extremely important; and it is also very helpful and informative – as I hope should be obvious by now.

So the lesson here, as far as moisture meters are concerned, is to use them with care, and *never* use the short pins that come with them as standard. *Always* spend that bit of extra cash and buy the hammer probe accessory: then use that instead. Or, of course, you could just guess at the moisture content – as most of the professional users of wood seem to do – and thus save yourself the cost of a meter and its accessories in the first place … (I'm kidding).

Now that I've told you about the two reliable ways of measuring its moisture content, I need to go back to thinking about the uses of timber in service, once it has been 'dried'. Then I can take a look at the actual methods of drying, in a bit more detail, to finish off with. And those methods are air drying and kiln drying.

'But hang on a minute,' I hear you saying: 'if wood has been kiln-dried, then surely it should be totally dry anyway – so there can't be any moisture left to measure, can there?' Wrong! The only time that wood is ever 'dry' (i.e. with no water left inside it at all) is when it is at, or above, 100^0C (such as when it's been inside the oven that's used to dry out the samples for calculating moisture content). And of course timber is never at 100^0C in normal use, is it? Which means that it must *always* contain *some* moisture, even when it is 'dry' enough to use in a building. So please be careful: just because a piece of timber might *feel* dry, that doesn't mean it *is* dry! And even if it is 'dry' it may not be dry *enough* to actually use for a specific job.

3.3   'Wet' or 'dry'? In-service moisture contents and 'EMC'

I posed a question earlier on in this chapter, asking whether a particular moisture content meant that timber was 'dry' or 'wet' when it came to a specific end-use. I asked if it is OK when it's at 20 % mc, or at 15 % mc, or whatever. Just exactly what is a 'suitable' moisture content to work with, when using timber and wood-based products? And is one particular level of mc going to be correct, wherever you may use that timber?

As always – it's not that simple! In addition to just measuring the *present* mc of any piece of timber; either one you are now using, or one that you intend to use in a certain location, you will also need to consider the effects of that location upon the likely *long-term* moisture content that the timber will achieve, once it has 'settled down' in its final place of use. That is because wood as a material has another somewhat unusual property: it is *hygroscopic*. That means it interacts with the moisture in the atmosphere of its surroundings: and so wood either gains or loses moisture, depending upon the relative humidity (RH) of the surrounding atmosphere and its own current level of inherent moisture (its mc). 'Wet' wood will lose moisture, by evaporation from its surface area, into a dry atmosphere; whereas 'dry' wood will take up moisture from a damp atmosphere, by the physical process of 'adsorption' of H_2O molecules into all of its exposed surface area.

This process of moisture gain or moisture loss to and from the wood is happening constantly. Each time the conditions surrounding the wood alter significantly – up or down – then the wood itself will alter its inherent mc to adjust itself up or down as necessary: until it once more gets into balance with its surroundings.

3.4   EMC

We call this 'balance' point its *Equilibrium Moisture Content* – or 'EMC' for short. And it should now – I hope – be fairly obvious that the desired EMC of any piece of timber (or any wood-based product or wooden component) will need to be established, for each location that you are intending to use it in. So, for example, the EMC required for a timber deck out of doors will need to be somewhat higher than the EMC required for skirting boards in a centrally-heated room; and so on.

3.5   Specification of desired moisture content

So how is the 'desired' EMC for any particular use established? The answer is that lots and lots of tests have been conducted by various Forest Products Laboratories around the world, on lots and lots of different pieces of timber, and in a large number of different relative humidities and temperatures. Then from that mass of data, some very detailed tables or graphs have been created, which can then help us to know what the effects of any atmospheric changes will be, on wood used under different conditions. And there is a particular set of EMCs for timber used in a UK context, in different surroundings (see Figure 3.2).

Many British and European Standards can help us with the specification of reasonably accurate EMC levels – that is, specific mc figures – that should be used for timber and wood-based products in different locations. One of the most helpful of these is BS EN 942: 'Specification for Timber in Joinery'. In something called the National Annex to this document – that's a special extra part that was put in just for us in the UK, at the back of that Standard – there is a Table, showing some recommended moisture contents for timber in different categories of use. These are: out of doors, indoors – either with or without heating – and so on (see Table 3.1).

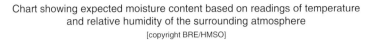

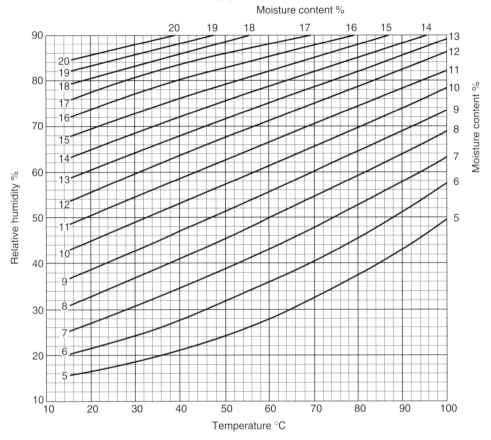

Figure 3.2   Chart showing expected moisture content (with permission)

Table 3.1   Expected equilibrium moisture contents for different end uses

Equilibrium moisture content %	End use
18	Carcassing timber
15 to 18	Exterior joinery
12 to 15	Interior joinery
12	Wood block flooring
10	Timber in permanently heated conditions
8 to 9	Radiator surrounds and timber over under-floor heating

So the real answer to the question; 'Is such-and-such a level of mc too high or too low?' has to be given in the form of my oft-repeated phrase: 'It all depends …'

It depends upon what the actual mc level of the timber is at the time you measure it; and it also depends upon where you plan to use the timber, once it is in service. But it also depends very much upon whether you expect those conditions to change in the future: such as the different atmospheric conditions between Summer and Winter. Or the change that will most likely happen when installing central heating into a previously unheated building. Or by delivering timber to a construction site, where it is initially exposed to external conditions; but it will eventually become the woodwork within the finished building: and that will involve it in all sorts of (not always helpful) adjustment in its moisture content, until it reaches its 'in-service' EMC. Basically, anything and everything that could reasonably be expected to change in the expected service life of a timber component should be considered – and then all of the possible effects of those changes, in respect of the timber's condition, must then be assessed and suitable steps taken to ensure that problems are minimised. To do *that* properly will require some additional knowledge about wood and its responses to moisture.

So, with all that in mind, there are some other properties of wood which you need to be aware of, with regard to its reaction to moisture: and two of these are the concepts of 'shrinkage' and 'movement' (which I touched on a little bit in Chapter 1, in connection with grain – but I didn't fully explain it at the time). First of all, though, I need to give you another small but important bit of extra information that you should know about the behaviour of wood with water inside it.

3.6   Fibre saturation point

A freshly-felled tree can easily be at a moisture content in excess of 150%. 150%! – now at first sight, this statistic looks slightly crazy: how can something be more than 100% of anything? But if you remember, mc is expressed as a percentage of the weight of the water it contains, *compared to the oven dry weight of the wood tissue* – so that means it is perfectly possible for the wood to hold one-and-a-half times more water *by weight,* than the wood itself weighs! If you look back at the example I used before, of a chunk of wood that weighed 1 kg when fully dry; then it would weigh at least 2.5 kg at the time it was freshly-felled. That total weight would be made up of 1 kg of wood, plus another 1.5 kg of water: therefore making it 150% mc.

The reason why wood can hold so much water is because – as I said in Chapter 2 – it is made up of cells that are essentially tubes with a hole down their middle (the 'lumen'): and these holes are capable of holding huge amounts of liquid water, in the form of sap. But the cell walls can also hold water; although in this case, it is held in the form of molecules of H_2O, that effectively 'muscle in' amongst the molecules of cellulose.

When the felled tree trunk begins to dry out at the sawmill – first of all whilst it is in storage in the log yard, and then during subsequent processing – the liquid form of water is the first kind of moisture that is lost: by evaporation into the atmosphere within the log yard or the mill buildings: and this speeds up as the bark is removed and the log is sawn into boards. As I have just said, that is a huge amount of water; so it takes some time to evaporate away (I'll put some timescales on drying, later in this chapter: but for now, I'll just stick to the principles).

It is not until the wood has dried down to about 28 % mc (a very long way from where it started, at around 150 % +) that we can say it has lost all of its liquid water, and that the cell lumens are now effectively 'empty'. But at this stage, the cell walls themselves are still full of water molecules, which are still, so to speak, 'packing out' the cellulose molecules within those cell walls.

This very particular – and highly important – level of 28 % mc (which is a key stage in the drying of wood) is known as its 'Fibre Saturation Point' – or 'FSP'. And it is defined as the stage at which the cell cavities are effectively empty of any liquid water, although the cell walls are still completely full of molecular water.

3.7 Shrinkage

It is not until this 'magic' level of 28 % mc is reached, and the wood itself then carries on drying downwards, that the timber will begin to get smaller. That's because, as the molecular form of water removes itself from amongst the cellulose molecules, the cellulose then 'packs' itself closer together, to take up the space which has been vacated by the water molecules. So this level of 28 % mc really is the point at which the timber actually begins to gets physically smaller – and this is state is what we Timber Technologists refer to as the phenomenon of 'shrinkage': that is, the *initial* reduction in size as the timber first of all dries out.

Please remember that timber *does not* shrink or expand at any moisture level above this 28 % mc figure. Of course, it gets *drier* as it comes down from its freshly-felled, wet state; and of course, it gets *lighter in weight* as well – but it does not get *smaller* until its mc starts to fall below the 28 % value of FSP. (I should just add, for the sake of accuracy, that 28 % is the 'accepted' figure for most wood species: some may achieve FSP at levels nearer to 30 % mc and others – notably 'Western Red Cedar' – will achieve FSP nearer to 25 % mc. WRC is also noted as being difficult to dry uniformly: it can have both wet patches *and* much lower mc patches in the same board.)

Oh, by the way, please don't forget that other very important characteristic of wood in respect of moisture changes, which I mentioned in Chapter 1, when I was telling you about the importance of grain direction. Remember: there is no appreciable dimensional change in the *length* of a 'normal' (i.e. well grown) piece of timber: it is given in most textbooks as being about 0.1 % from 'green' to 'seasoned' – and don't worry, I'll cover those terms in this chapter. So, for all practical purposes when dealing with timber in everyday situations, it is most important to understand that the shrinkage associated with drying causes a reduction in dimension only *across* the grain and not *along* the grain of the wood; and hence also it happens only across, and never along, the piece(s) of timber that you are using. But then, if wood's initial moisture loss causes 'shrinkage', what exactly is 'movement'?

3.8   Movement

I referred fairly briefly in Chapter 1 to the fact that when timber undergoes changes in its moisture content – either up or down – it swells or shrinks accordingly. And in this chapter, I've told you that because wood is hygroscopic it will always be prone to reacting with any atmosphere – either dry or humid – that it finds itself in. And those day-to-day surroundings may sometimes cause the wood to lose moisture; and at other times they may cause it to gain moisture: depending upon the wood's own inherent mc at the time.

Of course – as you now know – for every loss of moisture there will be a necessary reduction in the timber's cross-grain dimension (for the reasons explained above); and the reverse is also true. For every increase in moisture, there will be a necessary increase in the timber's cross-grain dimension as well. And it is these 'in-service' dimensional changes which result from moisture loses or gains due to seasonal or heating changes, and so on, that we should refer to properly as 'movement'.

Most importantly, you should also remember that movement *only* happens once wood is in service: so it generally only happens after a timber component has been installed. So it is not like shrinkage, which normally happens only once, at the time when the timber is first drying out after having been initially processed: such as in a sawmill.

Unlike shrinkage, movement is happening continuously: for the whole lifespan of the timber. It does not stop, even when the wood is supposedly 'dry'. Yes, it's true: even wood that has been in a building for 100 years or more, will

Table 3.2   Various timbers showing different movement classes

Movement class	Small movement	Medium movement	Large movement
Softwoods	Douglas fir Western Red Cedar	European Redwood European Whitewood	
Hardwoods	African Mahogany American Mahogany Iroko Jelutong Teak	Ash Birch Sapele Sycamore	Beech Ekki Karri Ramin

still 'move' if that building undergoes some significant alteration in its atmospheric conditions; such as having central heating installed. That's the thing about timber. It moves: and you need to make allowances for it.

I'll get on to the whole business of timber drying in a short while: but I just need to make another important point first. Because I need to tell you two very important things in relation to movement.

The first thing I need to say here is that movement – whilst of course being confined only to the cross-grain direction – varies according to *which* cross-grain direction we're talking about. As a general rule, movement in the radial direction (see Chapter 1) is only half the amount of that seen in the tangential direction, for the same loss or gain of moisture. Why this is so is not fully understood: but the general consensus is that it is because of the restricting influence exerted by the rays, which serve to 'hold back' the movement of the wood cells as they try to swell or shrink. And that restraining influence is absent in the tangential direction, since that is at right-angles to the direction of the lines of ray cells.

The second thing that I need to tell you is that the amount of dimensional change due to moisture movement is not the same for every species. For the exact same increase or decrease in moisture content, some timbers manage to move by relatively slight amounts; whereas others move considerably more; and others move by a very large amount indeed. We really don't know the precise mechanism by which different timbers can manage to do this: but we *do* need to know which timbers do which, in terms of amounts of movement, when we're using them (see Table 3.2).

Most reference books on timbers will list all or most of their various properties; and movement is no exception. The different timbers will be listed according to whether they have a 'small', 'medium' or 'large' movement. These terms are quite rough-and-ready distinctions, but they can still be very helpful in deciding which timber to choose for a particular job, where large or frequent atmospheric (and therefore EMC) changes are expected to occur in service.

Now I'll tell you about the ways in which we can hope to get timber 'dry'.

3.9   Kiln drying

The very basic term 'kiln dried', in relation to timber, is highly misleading: and in my view it may sometimes even be dangerously misleading. All too often, declaring that some timber is 'kiln dried' – but without giving any further details – leads to some false assumptions being made about it: by far too many of the professional users of wood. There is, so to speak, a doubly-damning delusion in the minds of far too many people, who speak glibly about 'kiln dried' timber, without realising that not only are they not giving enough information to help anybody in any way; but they are also somehow instilling into people's minds a false sense of security about the behaviour of that timber. First of all, the uninitiated believe that so-called 'kiln dried' timber is 'dry'

Picture courtesy of BSW

Figure 3.3a   A typical drying kiln (photo courtesy of BSW Timber plc)

(whatever that may mean!). And secondly, they believe that the timber is somehow dimensionally fixed and stable for evermore. But on both counts generally, nothing could be further from the truth.

And what, when all is said and done, *is* kiln drying, anyway? In reality, a kiln is not much more than a warm, damp shed, which is somehow temperature and humidity controlled, so that it can get gradually hotter and drier inside its four walls (see Figures 3.3a and 3.3b). The process of kiln drying – put as simply as possible – is that 'wet' timber is placed inside a kiln chamber and is then exposed initially to some fairly gentle warmth, along with a reasonably high level of relative humidity (RH). This ensures that the wood cells do not lose their moisture too quickly and thus create all sorts of quality and performance problems with the timber later on in use. As its time inside the kiln chamber goes on, the heat is gradually turned up and the humidity is gently lowered, in order to slowly encourage the moisture within the timber to migrate to the surface and then to evaporate away: but not *too* slowly, or the whole process will cost too much!

Thus the 'art' (and it is more of an art than a science, believe me) of good kiln drying is to balance all of the factors above – temperature, RH and the starting and finishing moisture content of the timber – to become part of an acceptable timescale, and without 'overcooking' the wood and thus wasting or seriously reducing its value. (This book is not the place in which to find huge amounts of information about kiln-drying techniques and the possible problems or solutions to bad drying: for that, see the Bibliography for some recommended further reading.)

It should be understood that the entire kiln drying process can be interrupted – or even stopped – at any point, depending upon the desired end result. Or, perhaps more mundanely, depending upon the time available for drying as

Picture courtesy of BSW

Figure 3.3b   Timber correctly stacked and awaiting drying in a kiln (photo courtesy of BSW Timber plc)

dictated by costs or production demands. Thus, the timber that eventually comes out of the kiln chamber will only be as 'dry' as the time and the process parameters up to that point will allow it to be. It may still not be 'dry' enough to actually use – so the term 'kiln dried' on its own, is essentially meaningless!

Without knowing which kiln 'schedule' (as the various drying regimes, temperatures and times are known) has been used, nobody can possibly say how 'dry' the timber is when it finally comes out of the kiln. It *may* be dry enough to use straight away; but all too often, it isn't – as I will demonstrate later on in this chapter. But I need you to realise that the term 'kiln dried', used just on its own, really only means that the timber has been in a warm, moist place for a while; and therefore it may possibly have lost some of its inherent moisture, if you're lucky. But that isn't really good enough, I hope you'd agree?

So, if the term 'kiln dried' should not – as I've just established – be used on its own, then what else is missing? The answer to that is its final percentage moisture content value! Left to the variable mercies of a kiln, and without any final mc specification being given for it, the vague term 'kiln dried' could, in reality, mean a level of moisture remaining within the timber that is nowhere near its preferred, or desired, or (more importantly) recommended EMC level. In other words, as I warned you earlier, the timber's mc may be nowhere near to being something that is actually suitable for its final intended use.

Therefore, somebody, somewhere, really needs to specify some meaningful numbers for this timber, in terms of its moisture content, to sit alongside those weasel words 'kiln dried' – at least, if that term is to have any practical value whatsoever. And such a description needs to be something meaningful: such as, 'Joinery redwood, kiln dried to 12%±2% mc' for example. (By the way, I should also explain here that giving a single value for the mc in any specification is also pretty meaningless: since timber cannot realistically be dried to an absolute level – it can generally only be dried to within a reasonably close range, which is normally reckoned to be about plus or minus 2 or 3% of a central figure).

Because the Timber Trade (or, if you'd rather, you can blame their customers instead: although someone has to take the blame for this!) don't usually want to pay anything extra for drying their timber, most commercial stocks – especially where Softwoods are concerned – are generally only dried down to something known – rather misleadingly – as 'Shipping Dry'.

But Shipping Dry is not really 'dry', as far as a normal end-use is concerned: since it is usually only somewhere in the region of 20%±2% mc. And at this moisture content, the timber is really only dry enough to allow it to be transported from the sawmill to the first purchaser, without it being excessively heavy (because water weighs quite a lot, remember): and also, with a bit of luck, without it going mouldy, or blue, or rotten.

Very helpfully, it is genuinely true that if timber is kept at a moisture content of 20% or lower, then it *cannot* rot. It may be at a very slight risk of a bit of mould or blue-staining, if it is held at around the 20% mc mark for a couple of weeks or longer: but wood will *never* go rotten, if it always kept at or below that 20% mc level. Never, not ever. And for that reason, you'll find that I generally refer to 20% mc as being the upper threshold of the 'decay safety limit' for wood.

But I was just talking about Shipping Dry and whether or not that can be regarded as being dry enough for all purposes. If you look at Table 3.1 you can decide for yourself whether 22% mc is 'dry' enough for most of the uses of timber in building (or inside buildings) in the UK. (Well, is 20% dry enough?)

Another crucial thing that people wrongly assume about kiln dried timber is that it will always remain stable in use: in other words, that it won't 'move'any more. But I explained earlier that wood is hygroscopic: meaning that it exchanges moisture with its surroundings all the time, until it reaches equilibrium. So even if so-called 'kiln dried' timber (i.e. wood that is currently Shipping Dry, at up to 22% mc) is stored outside in a wood yard or builder's storage area, it will still keep on drying down, because of its exposure to the outdoor atmosphere; until it reaches a new equilibrium that equates to the prevailing outdoor conditions. That EMC will usually be about 15–16% mc in the Summer and perhaps 17–19% mc in the Winter; in a UK context.

But let's say that someone has had the foresight to order *specially dried* timber: wood that has been dried down to the (quite low) level of 12% mc or even lower – plus or minus a suitable tolerance, of course! If that very dry timber is then stored unwrapped and unprotected in a yard – or indeed in an open-sided shed – then it *must* adjust its moisture content upwards, until it once more reaches an EMC that is in balance with the prevailing outdoor conditions. So, in either of the cases above – Shipping Dry, or 'extra dry' as we might call the latter – the timber simply *must* 'move' in response to those changes that are brought about by its adjustment to a new moisture content, as it finally reaches its EMC (by shrinking a bit, in the first example; and by swelling up a little, in the second case). So it should by now be obvious to you that the act of kiln drying does not confer any 'magical' immunity to wood, in respect of any future dimensional changes (i.e. kiln drying cannot, by itself, prevent any sort of movement).

3.10   Air drying

As the name sort of implies, this is the simplest method of drying wood: and one which relies only upon the naturally-prevailing atmospheric conditions to dry it. In fact, until relatively recently, it was the only method for drying timber, until the technology for steam kilns was developed in the 19th Century. But even up until about the mid-20th Century, air drying was the only method that was really needed for most of the uses of timber in construction. And why was that? Well, two words spring to mind: Central Heating. And I want to look at the effects which that has had upon our uses of wood over the past 60–70 years: but first of all, let me explain the rudiments of the air drying method.

Quite simply, all it needs is for planks of wood to be stacked in piles, with horizontal separators (usually called 'sticks', but in Scotland, also called 'pins') interleaved between the layers of boards, to allow the air to flow easily over all of the wood's surfaces (see Figure 3.4). And that's really about it. The sun and

Figure 3.4   Timber stacked for air drying – but not very well !

the rain provide the heat and the relative humidity in the air; and the wind, as it blows, moves the warmer, drier airflow over the timber stack, to encourage the evaporation of the wood's internal moisture.

Air drying is therefore reasonably simple to do and relatively cheap to carry out: but it has two major drawbacks. First of all, it is not very exact; so that a long, hot spell or a prolonged period of rainy weather can result in the wood drying either too quickly or to slowly – and which can then cause problems with its quality. And secondly, it has a much longer timescale to it. It really does take quite a long while to dry wood out of doors, without the mechanical processes and fine-tuning that you can get when using a chamber kiln. Before you ask: I do intend to give you an indication of drying times later, before I finish this chapter; but I'd really like to deal with all of the issues and the methodology first.

There is a third drawback to air drying: and – as I hinted just a few minutes ago – it is one which did not become apparent, until we in this country adopted central heating in a big way, in our homes, around the middle of the 20th Century. The EMC achieved by air drying (as you should expect) can *never* get below the Summer 'low' of about 15% mc (and that is quite logical when you think about it – because the conditions of the outdoor atmosphere will create their own specific EMC). So, because of this fact, air drying on its own just can't dry the timber to a low enough level to cope with the very dry levels that timber will experience when subjected to central heating (and latterly, with air conditioning as well).

Indoor heated conditions can result in EMC levels as low as 10% mc – sometimes even even as low as 6% mc – and such EMC levels are not uncommon. So timber components really *need to be* kiln dried, in order to get them down to a low enough level to suit their expected EMC when used indoors in our modern building conditions, without causing any deterioration problems such as shrinkage, distortion or cracking.

3.11 Timescales for drying timber

I promised to talk about this: so how long does it take to dry timber by the two basic methods? The answer is not *quite* 'How long is a piece of string?' – but it's not very far off it!

The problem is that there are a lot of different factors involved. First of all, what is the starting mc of the timber to be dried? And then, what level do you need the timber to be at, when it has finished drying? (The bottom limit of mc that is imposed by air drying notwithstanding.) Also, how thin or how thick is the timber? Thicker timber takes longer to lose its moisture than thinner timber (which is logical when you think about it; since there is a greater distance for the moisture to migrate up to the surface). Of course, thicker timber is more likely to contain more moisture in it anyway, since there's more wood substance there in the first place. And another thing: which species of wood are you trying to dry? As a general rule, Hardwoods will take longer to dry than Softwoods:

and some timbers are a real problem to dry quickly (the term we use is 'refractory'), and so these wood species will need more time in the kiln – or in the air – anyway.

So now, assuming that we've unscrambled most or all of the factors above, and reduced the variables to manageable proportions: how long will it take to get your wood dry? Well, as I said earlier, this is not the book in which to go into huge detail on drying: but as a guide, I will say this much, at least.

To dry fresh-sawn Softwoods in a kiln at a sawmill, down to 'Shipping Dry' (around 20% mc or so, if you remember), will take from two to four days, depending upon the thickness of the boards. But to dry the same timber, in stacks, by air drying, could take four or five *months*. For many Hardwoods (which in the UK are often dried wholly, or at least partially, by air drying) the timescale is very approximately one year per inch thickness, to get down to the 'air dried' level (that is, 15–17% mc depending upon the time of year). Thus to dry 1-inch thick Oak planks would take a year to get from 'green' – that is, fresh-sawn – to about 16%; and 2-inch thick planks would take about two years to achieve the same result.

Hardwoods for special uses – such as furniture (which of course, should really have an indoor EMC to avoid all sort of problems) – will often need to be finished off in a kiln, even after a period of air drying, in order to achieve that bit of additional drying that is required to lower their EMC to the desired level: typically 10–12% mc. And that extra drying may be about another week, give or take a day or two, depending upon species and thickness – of course.

So you see: drying timber is not by any means an 'accidental' process. Or at least, it shouldn't be!

But all too often it is, in effect, accidental: because nobody has actually had the foresight, or the time, to do it right in the first place.

Now I'd better summarise what I've told you so far in this chapter, about Water in Wood.

3.12   Chapter summary

First of all, I defined what 'moisture content' actually means: that is, the amount of water in wood *by weight*, compared with its completely dry weight, as wood tissue alone. And I used the abbreviation 'mc' to refer always to 'moisture content.

I showed you that, although we can check the mc of any bit of wood accurately by using the 'oven dry' method; in normal, everyday usage, you would be much better off using a properly-calibrated Moisture Meter – but just so long as you learn to use it properly, and therefore avoid fooling yourself (and others) with incorrect or inaccurate readings.

Next, I looked at the correct specification of moisture content for timber in various conditions of use: the so-called 'in-service' moisture contents. And I gave you the concept of 'equilibrium moisture content' – or EMC – as a guide as to how 'dry' the timber needs to be, to do any particular job properly. I also mentioned the Timber Trade's notion of 'Shipping Dry' and asked whether that was always 'dry' enough for use. I told you helpfully the 'decay safety limit' for wood: which is 20 % mc or lower.

Then I explained the main methods of drying timber – kiln drying and air drying – with their pluses and minuses. And finally, I looked at how long it should take to dry wood properly. Anything from several years for air dried Hardwoods; to just a few days for kiln-drying Softwoods (although they will not by then be fully 'dry' – they will only be dried down low enough to be able to be transported, without going blue or mouldy).

And now that I've covered the basics of wood's properties and behaviour, in response to water, I'd better help you sort out some of the more basic the issues surrounding how to specify it correctly.

4 Specifying Timber – for Indoor or Outdoor Uses

In the previous chapter, I outlined the basic rules regarding moisture content in timber components used in various locations. But simply getting the moisture content of the wood right – vital though that is – is only a relatively small part of the story. You will also need to bring in some of the basic stuff on wood properties that I talked about in the first couple of chapters, in order to deal properly with using different wood species in different places. You'll need to do more if you intend to get the best possible performance out of the wood you want to use.

And to help you get to grips with the various things that you should know, I'm now going to introduce you to a couple of very useful European Standards. These in turn will give you some pretty important factors to take in to account, when looking at the uses of timber under different conditions. But just before I tell you about these particular Standards that I want to introduce you to, it's worth having another slight digression, to explain the status of the various Standards that you might come across.

4.1 British and European standards

Many of our British Standards have been superseded over the years, by a number of new European Standards. These European Standards are published as 'EN's' – that is, Euro Norms – and they are published by the European Union in the three Official EU Languages of French, German and English. (Which is quite fortunate for us Brits, who aren't very good at interpreting any foreign stuff!) And the rules of the EU state that any European Standard – on any topic – which covers the same ground as any EU country's own National Standard (such as one of our British Standards), must then replace that National Standard. It is all done in the spirit of 'harmonisation' of regulations: and surely that can only be a good thing, whatever you may think of the rest of the EU's penchant for bureaucracy.

Wood in Construction: How to Avoid Costly Mistakes, First Edition. Jim Coulson.
© 2012 John Wiley & Sons, Ltd. Published 2012 by John Wiley & Sons, Ltd.

One other thing about ENs is that, in all EU member states, each of the Standards will include, as part of their contents, both their Europe-wide scope (the 'essential rules') and also their particular National application (the 'local rules'). Thus in the UK, these ENs are published as 'BS ENs': and whenever there is something specific to the UK, you will always find a 'National Annex' at the back of the Standard, detailing the stuff that is only applicable to the UK context.

Now I need to get back to the things that I think you need to know about the uses of timber in various places: which is the main thrust of this chapter. And it will be helpful to remind you of the particular properties of wood that are especially applicable in this context.

4.2   Durability and treatability of different wood species

If you cast your mind back, you should recall that I explained about the Natural Durability of timber; and how it is closely tied up with the particular nature of the heartwood of any particular type of tree. And I also explained that the microscopic cell structure of wood can either help with, or hinder, its uptake of liquids.

Bearing these two fundamental properties in mind, I can now tell you that there is a very helpful European Standard available for us to refer to, on these very topics: and it is BS EN 350:1994. This particular EN deals fairly comprehensively with the whole issue of the durability of certain wood species and their ease or otherwise of being impregnated with wood preservatives: and it then relates these properties to exactly *where* the wood is being used (see Tables 4.1 and 4.2). Part 1 of BS EN 350 gives the method for testing the Natural

Table 4.1   Table of Natural Durability ratings from BS EN 350-1: 1994

Class	1	2	3	4	5
Durability Rating	Very Durable	Durable	Moderately Durable	Slightly Durable	Not Durable

Table 4.2   Examples of timbers within each durability class

Class	1	2	3	4	5
Softwoods		Western Red Cedar Yew	Pitch Pine	European Whitewood Western Hemlock	
Hardwoods	Afrormosia Greenheart Teak	Ekki European Oak American Mahogany	African Mahogany Sapele	Elm Red Oak	Ash Birch Sycamore

Durability of wood species, along with a guide to the different Durability classifications. But it is Part 2 of BS EN 350 which, most helpfully, lists both the Natural Durability and treatability characteristics of a number of selected species of timber used in Europe.

By using this very helpful Standard – and also just as importantly, by making sure that you reference it when creating any formalised Specification – it will be possible to select a timber, and/or a preservative treatment, that will be suitable for use in various locations. And then, the exact nature of, and the conditions that relate to, those locations – in the context of timber use – are given by the second of the two helpful Standards that I said I would tell you about: BS EN 335.

4.3   Use classes

This second Standard of my 'helpful pairing' also has two distinct Parts to it. BS EN 335-1 considers only the performance of preservative-treated solid wood; whereas BS EN 335-2 deals only with the performance of untreated solid wood, which has its own naturally durable heartwood.

This Standard then sets out five basic classifications – which are effectively, situations or circumstances of the use of wood – where timber components are employed. And it calls these situations, quite unambiguously, 'Use Classes' (see Table 4.3). (Just for clarity, I ought to tell you that, in a previous version of the same Standard, these classifications were known as 'Hazard Classes'; but it was thought that the term sounded a bit too 'risky – and so the classifications were changed, to become the much more friendly and tame-sounding – and perhaps more socially-acceptable? – Use Classes instead.)

In the intervening years since the first of the 'Use Class' versions of EN 335 was published, there has been a sort of 'fine tuning' of some of the specific Classes – most notably with the help of the Wood Protection Association – and so you'll find that some of the classifications now sport a couple of sub-divisions

Table 4.3   Simplified version of table showing use classes to BS EN 335-1

Use class	General service conditions	Exposure to wetting during service
1	Above ground, covered and dry	None
2	Above ground, covered with risk of wetting	Occasional
3	Above ground, not covered	Frequent
4	In contact with ground or fresh water	Permanent
5	In salt water	Permanent

to them. However, I personally think that the original five Classes are still quite good enough on their own, if used properly, to give the sort of helpful guidance that specifiers need. And if you follow that guidance properly, with a little thought behind it, you won't go far wrong.

4.4   Examples of timbers employed in different use classes

I think that it would now be very useful to consider a number of examples of the sorts of real-life places where timber could find itself used; and to suggest some practical examples of wood species and/or wood species-plus-wood preservative combinations that could be used in such different locations and under the different atmospheric or moisture conditions which will then prevail. And I would like to do that in a logical sequence: going from the 'safest' use to the most 'hazardous' (there's that word again!) conditions of use.

Before examining each of the five Use Classes in some detail though, I'd just like to caution you about the words 'hazard' and 'risk': which are two words that many people blithely use as though they are completely interchangeable in meaning; but which of course, they are not.

4.5   Hazard and risk – and their relative importance

A 'hazard' – as you may guess from the nature of the word – is something that can, under the right (or wrong!) circumstances, cause a problem. And in the particular context of the uses of timber in construction, one of the biggest hazards is moisture. Problems can result if there is too much or too little in the surroundings; or too much or too little in the wood itself, and so on. But, although there may be some sort of theoretical hazard lurking around because of the particular *way* in which a timber may be used, or because of *where* it may be used, or even *how long* it may be used in that situation; there is also a need to evaluate the *likelihood* of that hazard having any major effect on our particular timber in service.

And the degree of likelihood of any problem actually developing, is what we mean by 'risk', when we actually use a particular timber in a particular place. Think of it this way: although something *might* possibly happen, if it is highly *un*likely to happen in reality, then the risk associated with it is actually very low – despite the potentially damaging nature of the particular hazard involved: such as rot, perhaps. Conversely, if the nature of the hazard is fairly minor (a small amount of shrinkage, say), but the likelihood of its occurrence is very strong, then we would say that the risk of any timber being thus affected is great; and we might wish to do something to try and avoid it, even if its effects might be less of a problem to us.

So, as you read about the examples of the different Use Classes below, please bear in mind what I've just said about the inter-relationship between 'hazard' and 'risk': and that should then help you to decide what to do under different circumstances, in case of any doubt. And you may also care to recall what I said in the Foreword: that it is *never the fault of the wood*! Please read on.

4.6   Use Class 1 – examples

This is, in a way, arguably the 'best' or 'safest' environment in which to use timber. But it also has things about it which you need to pay attention to, if you are still to avoid problems: or at least, if you are to try and minimise any problems in service.

Use Class 1 relates to those situations where the timber is always going to be used in a 'dry' atmosphere. And that is one in which – according to BS EN 335-2 – any timber will have a moisture content that can never exceed 20 % for the whole of its service life: '… such that the risk of moulds, stains or wood-destroying fungi is insignificant.' So far, so good.

Typical locations where Use Class 1 applies would include the interiors of most domestic houses, or offices, or shops, or hotels, or retirement homes (see Figure 4.1). But each of these different places – whilst being 'dry' within the very wide parameters of that vague term – has its potential problems and pitfalls. And these pitfalls for the unwary specifier or user relate not to any likely risk of the timber developing blue stain, or going rotten during its lifetime: far from it. Instead, they relate to the proper specification and maintenance of a

Figure 4.1   Typical Use Class 1 location – built-in fitments in a bedroom

correct moisture content within the timber (its EMC, no less), in order to reduce the likelihood of potential problems due to movement or shrinkage. In other words, to try to avoid cracking and distortion, or creating unsightly gaps between components after they've been installed.

I explained in the last chapter that wood reaches a different EMC, depending partly upon where it is used, but also upon *when* it is used: in Summer, Winter, and so on. Although the 'when' is less important with indoor uses, it can still have an influence, because (for example) the windows of the property may be open and its central heating turned off in the Summer months; whereas the exact opposite is far more likely to be the case in the Winter months. And that potential variation in atmospheric conditions needs to be at least considered, when specifying an appropriate end-use moisture content for timber components being delivered for any job – either in new works, or in refurbishment or restoration projects. And that moisture content may be different – or it may have tighter or looser tolerances – depending upon whether it is structural timber or joinery timber, or veneered items, such as flush doors.

For example: it would be a waste of both time and money (in kiln drying energy-cost terms) to specify a moisture content as low as 8–10 % for joinery timber to be delivered to a site in the Summer, when open windows and a lack of any artificial heating could mean that the real EMC would be nearer to 12–14 % at the time of its installation. But delivering air-dried timber, such as whitewood floorboards, at around 16–18 % mc in the Winter to an existing building, when the heating is likely to be turned up high, and any ventilation to the outside air is likely to be nonexistent, could be disastrous: since the required EMC would then be down to about 10–12 %, or maybe even lower. In the first example, some swelling of the timber components would result; and in the second example, a degree of shrinkage and possibly also some distortion and cracking, would be the most likely outcome: in either case, it is a problem that could easily be avoided with a little forethought.

But it's not only the time of year that is important: the state of completeness of the building is also a vital factor to take in to consideration. Thus a new-build scenario, where wet plaster has just been applied, is not exactly the ideal location in which to fix architraves at 10–12 % mc; since the RH of the surroundings is bound to be raised considerably by the evaporation of moisture from the wet plaster into the atmosphere of the building – into the very same atmosphere that now surrounds the timber! However, delivering air-dried or 'Shipping Dry' floor joists at around 20 % mc or just below, to a partially-completed building site, is probably going to be reasonably OK; because the timber will have time to 'settle in' as the building progresses towards completion. (That is, provided it's not midwinter when Final Completion happens, and the heating is then wound up to maximum, to 'help dry the building out' before it's occupied!).

Finally in this particular bit of the Use Classes section, I have to say that the *sort* of building in which the timber gets used is most important, too. A barn or a storage building will be 'dry' inside: but it will probably develop an EMC of no

Figure 4.2   Cracking in skirting caused by over-drying in situ

more than about 14–18 %, depending upon the season, since there is no heating within it. Therefore a sensible level for the 'as delivered' mc in timber components for such an end use would be air-dried or equivalent: at about 16 % ± 2 %.

But at the other end of the scale – say in an old peoples' home (or 'retirement home', to use the more PC term these days) – the heating is usually running constantly, even in warmer weather, and so the indoor temperature is always very high and the RH is usually extremely low: resulting in an EMC in the timber of about 6–8 %. Therefore, delivering air-dried timber here – as happens far too often in my experience – will be disastrous. The result will almost inevitably be severe shrinkage, leading to cracking and also perhaps distortion in the components (see Figure 4.2). This can also happen to the faces of veneered doors too; where excessively low RH can cause slight shrinkage, even in a very dry veneer – just enough to open up any 'checks' that were created from the peeling of the veneer. (See the chapter on sheet materials for a further explanation of how and why these checks can happen). And it's not only the wood that suffers: the old folks get quite dried out too, with dry eyes and throat infections being all too common complaints in these sorts of places – but that's another story …

So, although Use Class 1 is intended to be for 'dry' timber (that is, wood which is always kept below 20 % mc), you will still need to know what version of 'dry' you are dealing with, and then adjust your specification – and the 'as delivered' mc – accordingly, if you are to avoid or at least minimise some unsightly 'cosmetic' problems in service.

4.7 Use Class 2 – examples

Although this is a theoretically higher level of risk, it is still a reasonably 'safe' Use Class for timber: although it is not quite as guaranteed to be free from staining and decay problems as is Use Class 1. In Use Class 2, the timber will be mostly dry for most of its service life; but there can be times when, in the words of BS EN 335-2: 'the moisture content of solid wood occasionally exceeds 20 %, either in the whole or only in part of the component and thus allows attack by wood-destroying fungi.' So when might that happen?

An example of Use Class 2 would be timbers in roof trusses above a swimming pool, where condensation could occur during certain times of the year; or during certain heating cycles; or perhaps if there are insufficient air-changes happening within the building. Closed roof voids – both in pitched roofs and (especially) in flat ones – are another 'risk' area; where insulation may block ventilation paths; or where slipped slates or other damaged roof coverings may not be quickly repaired, thus letting in rainwater over a long period of time. This leakage can then run down or through the roof timbers, and gather in one location – typically at rafter ends – and can thus raise the mc of the timbers, in one particular location, above that all-important 20 % level, which of course is the decay-risk threshold (see Figure 4.3).

In these two examples, and in many others you may think of which could fit the application of Use Class 2, there is usually some specific feature of the

Figure 4.3   Leak in roof, causing dampness which may lead to decay

building's design or maintenance regime which, if not properly thought through or carried out in practice, could lead to occasional or periodic rises in the mc of the timber: either generally or – more usually – quite locally. So, apart from needing to specify an appropriate moisture content at the start of the contract (appropriate, that is, to the 'as built' EMC of the structure in service) the designer or building manager (or whoever it is who takes responsibility for its future maintenance and repair) will need to consider the likely risk of a future rise above that crucial decay threshold of 20 % mc. He or she will then need to weigh that risk against the cost of any future maintenance, or any remedial or repair works that may become necessary. Remember the discussion on 'hazard' and 'risk' a little earlier in this chapter? The hazard of raised mc levels, leading to some decay, is a genuine possibility in Use Class 2: but how likely is the risk of that hazard occurring? And what is the likely cost of putting the damaged timber right again?

If the person responsible for the building's maintenance thinks that the risk of potential future decay is a significant one, or if the costs of any likely future repairs will be disproportionately high, then some other strategy will be needed, apart from purely getting the 'as delivered' mc as near as possible to the 'in service' EMC.

For example: if the cost of a roof repair in a high-rise building would be prohibitively expensive in the future – most likely, because of the need for scaffolding or other access needed to dismantle large areas of the roof – then it would be sensible to specify a preservative treatment to the roof timbers at the start of the building contract, so as to minimise the risk of decay, should any leaks in that roof happen in the future. And although such preservative treatment is an extra capital cost, it is generally far less money to find 'up front' than it would be to fund the location and repair of an unknown amount of decay damage later on.

In the swimming pool roof example I gave earlier, this could also be dealt with by means of some type of proprietary preservative treatment. But that is not always appropriate – or necessary – since the timbers in such roofs are very often visible: and so these roofs lend themselves to being built in some timber other than the usual 'cheaper' Softwood alternative. Therefore, the selection of an attractive – but also structural – Hardwood with an appropriate degree of natural durability (that is, Class 3 or better from BS EN 350 – see Table 4.1 above) could be an elegant way to cope with the need for some sort of 'insurance' against possible future decay risk.

When choosing a suitable Hardwood, the range of possibilities may seem endless: but in reality, price and availability are going to restrict your choice somewhat. Oak (*Quercus spp*), Iroko (*Milicia excesla*), or even Opepe (*Nauclea diderrichii*) are all fairly likely candidates here – along with at least ten others: should the designers wish to be more adventurous in their choice of timbers that are both decorative and structural.

Sometimes, the decision about the need for treatment has already been made for you. Such as the general requirement (as stipulated by, amongst others, the National House Building Council) that all of the vertical studs within an external timber frame wall panel should be treated with preservative, as a precaution

Figure 4.4   Studs and rails in a timber frame wall – possibly at risk from interstitial condensation

against the (albeit very low) risk of interstitial condensation: thus leading to a rise in the moisture content of those timber studs, which in turn could lead to some decay (see Figure 4.4). But more often than not, the decision about whether preservative treatment should be done will be one for the designer or the building contractor: and this decision needs to be made in the knowledge that a little extra cash spent now may save a huge repair bill later on.

4.8   Use Class 3 – examples

Both of the first two Use Classes belong, very firmly of course, to the 'indoor' variety of the possible uses of timber. But of course, wood gets used a tremendous amount out of doors too. And the most 'timber-friendly' of the potential outdoor uses are those where the moisture content of the wood will be not always be above 20 % for ever; but it could still be over that level for quite a bit

Figure 4.5a   Typical Use Class 3 location – fence rails

of the time. Or: '… frequently, and thus it will often be liable to attack by wood-destroying fungi,' as BS EN 335-2 puts it. But timber that finds itself in a Use Class 3 job will not be kept permanently wet all of the time: so this Use Class fits all of those cases where timber gets used in an external location, but where it is neither in direct contact with the ground, nor is it ever fully soaked in water.

Examples of such Use Class 3 situations are: boards or rails, as part of a fence (but *not* the posts – that's the next Use Class!); deck boards and other parts of a deck superstructure (but none of those components that actually touch the ground); external timber cladding; and all of the outdoor elements of doors, windows, cills and thresholds (see Figure 4.5a and 4.5b). In fact, Use Class 3 covers basically anything that is above the dpc level in a building, as well as anything that is not directly in contact with the ground, in all other situations of timber use out of doors.

Once again – just as in the example I gave you for that hypothetical swimming pool roof in Use Class 2 – your choice here is between using a preservatively-treated timber (generally some type of Softwood); or a more durable timber (which most often will be some species of Hardwood). If you opt for the latter, then you must of course specify that it is to have a Natural Durability rating of Class 3 or better (as given in BS EN 350): that is, a wood species whose heartwood is rated as Moderately Durable, at the very least.

It is an unfortunate coincidence that these two highly useful Standards – BS EN 350, covering Natural Durability and Treatability, and BS EN 335, covering

Figure 4.5b   Typical Use Class 3 location – deck boards

Use Classes – each lists five specific Classes within it. And in my example above, it is – perhaps a trifle confusingly – Class 3 from *both* standards which is the applicable level. I advised you that, if you do not opt for preservation treatment, then you will need 'Class 3' of BS EN 350 to ensure that you have a Moderately Durable timber: but if you *do* need to specify some treatment, then it must be stated as being to 'Use Class 3', when dealing with a timber of lower durability that is to be used out of doors, in an out of ground contact situation. Is all of that reasonably clear? I do hope so!

As with many of the 'ordinary' situations of timber use in the UK, the reality is that most external applications of timber for things like decking, cladding and fencing will use Softwoods: very probably because they are the cheaper option (see Figure 4.6). And with a very few exceptions (such as 'Western Red Cedar' (*Thuja plicata*) – which not a true Cedar, I should add, but which nonetheless has a good Natural Durability rating) pretty well all of the usual suspects amongst the Softwoods will need to be treated with some type of proprietary wood preservative, in order to achieve a performance suitable for Use Class 3. Even that last – and seemingly simple – statement is another thing to be very careful about.

Too many people – the majority of the Timber Trade included – seem to get by with just glibly talking about 'treated' timber. But that's simply not good enough, I'm afraid. Essentially, it's the same reasoning behind why asking for 'kiln dried timber' is not good enough either; as I told you in the previous

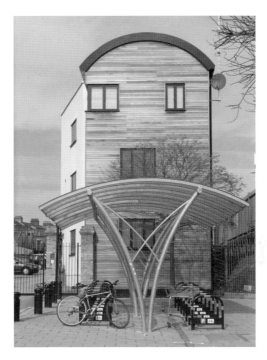

Figure 4.6   Western red cedar used as external cladding (photo courtesy of Canada Wood UK)

chapter. In the same way that the rather vague term 'kiln dried' could mean all sorts of possible levels of moisture being left inside the timber; so too can the equally vague term 'treated' mean that the timber might or might not contain enough preservative chemicals, at a good enough depth of penetration within the timber, to do the job required of it. So you will need to ask for the timber to be more than just 'treated', in order to make a less durable wood species do a proper job, for long enough.

So you must actually specify that timber is to be 'treated to Use Class 3' in order to get what you really should have, if that's what you really want. In other words, if you want to achieve a 'desired service life' of at least 15 years out of doors, away from ground or water contact, then you need to specify some preservatively-treated timber that has been correctly processed, by a reputable company, to meet UC3.

Of course, as I said earlier, you always can use any timber (be it a Softwood or a Hardwood) with a rating to its heartwood of Moderately Durable (or better). And the choice of such timbers for outdoor use is very wide: so, as I hinted earlier, it is cost that is generally the biggest influence on your final specification. For example: an ideal timber for use in external joinery would be Central American Mahogany (*Swietenia macrophylla*); although it can be quite tricky to get hold of these days, because of 'Green' concerns about its sourcing. But it is certainly a good choice in purely technical terms – in respect of its properties – since it is rated as Very Durable and it also has low movement characteristics, *and* as an added bonus, it is very easy to machine without any difficult grain problems. However, Iroko (*Milicia excelsa*) would make a very good alternative, as would Teak (*Tectona grandis*): and this latter timber is increasingly available these days as plantation-grown stocks from South Africa or a number of South American countries (such as Bolivia). As an added bonus, such plantation stocks often carry with them the additional cachet of 'sustainable' forest Certification and the increasingly popular Chain of Custody credentials – thus placating the 'Green' lobby and so removing another problem from the ever-more-complex wood supply chain.

A word or two of caution: be careful when choosing a timber such as Meranti (*Shorea spp*): because this is not just *one* single species of wood – instead, it comes from a range of up to 15 different species of the *Shorea* genus, from various parts of the Far East – and so not all of it will necessarily have the desired level of Natural Durability for a Use Class 3 job. Thus Meranti will require some preservative treatment before it is acceptable for use as, say, external joinery. (And it is stated as such in the National Annex to the European Standard on Timber in Joinery, BS EN 942.)

This last point harks back to what I told you in the very first chapter: that the term 'Hardwood' does not confer any magic properties onto a timber: and it certainly does not mean that any particular timber is suitable for everything – including use out of dors without treatment – just because it happens to be called a 'Hardwood'.

4.9   Use Class 4 – examples

Now you are getting into dangerous – or rather I should say, more hazardous – territory. Because this Class is another step further down the trail of those moisture-related 'hazards' that can beset the uses of timber, when it is employed in much wetter places. Most of these places where timber is subjected to high levels of decay risk are pretty much always out of doors: although, strictly speaking, Use Class 4 relates to all of those cases where timber: 'has a moisture content in excess of 20 % permanently and is liable to attack by wood-destroying fungi,' according to BS EN 335-2.

So effectively, the uses of any wood species in this situation will always result in some timber components – or at least, significant parts of them – becoming fully 'wet' all of the time. This can of course happen through direct immersion in water – such as with the legs of a jetty in a river or a lake; or with lock gates in a canal – or it can happen through the timber soaking up water from wet soil when it is embedded in the ground (see Figures 4.7a, 4.7b and 4.8).

This last situation is by far the most common way in which timber becomes permanently wet: in fact, direct contact with the ground is by far the greatest

Figure 4.7a   Typical Use Class 4 location – fence posts

Figure 4.7b   Typical Use Class 4 location – lock gates

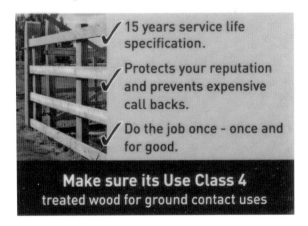

Figure 4.8   WPA brochure – Make Sure It's 4

cause of decay in most timber components the world over. (Of course, trees rot down when they fall to the forest floor: but I'm talking here of the times when we have deliberately chosen to put timber into harm's way).

The soil is frequently very wet when it is at or near to natural ground level (i.e. not well-drained or built-up above the water table); and of course, some types of soil are much more prone to retaining moisture than are others – heavy clays being particularly troublesome. Therefore, any timber component that is embedded in the soil will tend to draw moisture up out of the ground, along the grain of the wood: in exactly the same way as the wick in an oil lamp ('I'm not *that* old!' I hear you cry). However, once a substantial part of that timber component extends above ground, then evaporation will take place, transferring most of that moisture into the air, from the above-ground portion. So that element of the component will soon dry down again, to a level below 20% mc.

Conversely when it is completely buried, much deeper underground, that portion of the timber component which is surrounded by completely wet soil will become very, very wet: fully waterlogged, in fact – and so it is then sitting in an environment which is referred to as being 'anaerobic'. This means that there is effectively no oxygen available for any decay fungi to be able to live; and so the timber is effectively 'safe' when it is buried or used at a greater depth, below the water table.

However, the absolute ground level situation is (to use modern parlance) the so-called 'Goldilocks Zone': where everything is 'just right'. Just right to encourage rot, that is! There is just enough moisture (where the mc of the wood is somewhat above 20% and so the timber component is not too dry); just enough air (because the wood is not fully waterlogged); and also, a reasonable supply of food (provided of course by a susceptible wood species). That is why timber components, if they are going to do so, will *always* rot at the ground line – you only have to look closely at some fence posts or gate posts the next time you come across any, to see if I'm right. (I am) (see Figure 4.9).

Apart from fence posts and gate posts, other examples of timber components that are at severe risk of decay through ground contact are: the legs of any deck support structure (unless the particular design of the deck, plus some careful detailing, can avoid this situation – which indeed they can, with just a bit more thought); plus any other decking components which are resting upon or touching soil or low-lying vegetation. Indeed, any of the timber elements within a building which are located below the dpc level – such as the floor joists in a damp cellar – will be at risk of decay through long-term exposure to high levels of moisture (see Figure 4.10).

In exactly the same way as I explained to you previously when dealing with items exposed to Use Class 3, the way to avoid problems with any timber in Use Class 4 is either to treat it (to the right level!) with preservative, or to use a wood species of an appropriate Natural Durability classification. But even after having decided that one, you will still need to take a bit more care, to get everything just right …

The term: 'appropriate', when talking about a timber's Natural Durability rating for any application under Use Class 4, means at least Class 2 or better from BS EN 350. (The reason why it's Durability Class 2 is because, the higher you go up the scale, the more durable the wood species are.) So you will need to both find *and* specify a species of wood that is rated as either 'Durable' or 'Very Durable' for all of the timber components that are to be used somewhere where they will get wet and remain wet all of the time.

Alternatively, when considering the option of using preservatively treated timbers, you will need to think about the 'treatability' of any wood species that you intend to specify: because, as I said in the first chapter, not all timbers are the same, in terms of their individual properties. So, in order to satisfy Use Class 4 with a timber that is of low Natural Durability, you will need to use a timber that can be treated with preservative to an adequate depth of penetration and which leaves enough concentration of active ingredients within it to resist decay for the requisite length of time – which is generally a minimum of 15 years.

Figure 4.9   Post rotten at the ground line

A quick aside. These time periods that we expect timber components to survive for, are called 'desired service lives': and they were first introduced in BS 5589 in 1989. At this time, two separate periods were stated: 20 years and 40 years. A bit later on, in 2003, the newer British Standard on the preservative treatment of timber – BS 8417 – changed these periods to 15 years and 30 years; but it also added a new, longer period of 60 years. This last period was included in order to bring the expected lifetime of timber components into line with the expected minimum life-span of a permanent structure. (And in fact, the NHBC currently require any timber deck structure that is built as part of a new house, to have a designed service life of at least 60 years. And it can be done!)

But now back to Use Class 4 and the specification of a treated softwood. The best wood species to do this for you most easily and fairly cheaply, is any sort of Pine: that is to say, any species of the genus *Pinus*. That's because all of the 'true' Pines have a treatability rating that is described as 'easy to treat'. And our usual choice in the UK, is the timber that the Timber Trade will sell under the name of 'European redwood' – *Pinus sylvestris*.

Figure 4.10   Ground floor joists attacked by cellar fungus – *coniophora puteana* (photo courtesy of Bob Caille)

Conversely, any species of the genus *Picea* (true Spruce) is notoriously difficult to treat: since Spruce is described as being 'resistant to treatment'. This is because of its microscopic cell structure, as I explained in a fairly simplified way in Chapter 1; and so Spruce is not an ideal timber to use, where the higher loading of preservative treatment necessary to satisfy Use Class 4 is required. (Some recent work by the Wood Protection Association has shown that it is possible to achieve Use Class 4 with some air-dried Spruce, under certain conditions: but in general, timber treatment companies will not rush to give you a guarantee of much over ten years for treated Spruce timbers; and even then, only with additional provisos as to how you must look after it in service.)

Unfortunately for us who live and work here, the UK is somewhat 'blessed' with an abundance of relatively cheap, home-grown Spruce. It is mostly Sitka Spruce (*Picea sitchensis*) that is native to Western North America, but which has been planted in the UK by the Forestry Commission for over 80 years – and which most of the customers for 'cheap' wood seem to prefer to buy on price, rather than on performance. (Don't get me wrong: Spruce has some excellent characteristics: such as constructional strength; and it gives higher yields than Pine when being graded; but it is not an ideal timber to try and put a lot of preservative into, when you particularly need that ability for the higher-risk Use Classes, as I've been outlining here.) So what I'm saying is, treated softwood is fine for Use Class 4: but unless you can get someone to give you a guarantee for 15 or 30 years with Spruce; it's best to opt for Pine for your fence posts – and definitely for your 60-year-life decking supports.

4.10   Use Class 5 – examples

It's pretty unlikely that the majority of you will have to refer to this Use Class in the normal course of dealing with timber in construction, since it only involves contact with salt water. Of course, timber used in such a location will again be permanently over 20% mc – but as BS EN 335-2 says of the real danger: '… attack by invertebrate marine organisms is the principal problem.'

In the UK, the main culprits that eat wood at the seaside are the so-called 'Shipworm' (*Teredo spp*) and – I love this name! – the Gribble (*Limnoria spp*). These latter creatures bore small holes all over the surface area of any susceptible wood, which is immersed in the sea (or, to be more precise, the Gribble works in the inter-tidal zone: which is why the legs of jetties tend to wear thin in the middle – a phenomenon known as 'waisting'); and the damage that they can do is considerable (see Figure 4.11). But the Teredo is even worse: because it

Figure 4.11   Pile attacked by limnoria (gribble) (photo courtesy of Simon Cragg)

burrows along the entire length of a timber member until it has devoured it almost in its entirety (see Figure 4.12). I have seen a 10-metre-long jetty leg reduced to the appearance and consistency of a Swiss cheese!

The only two practical answers in respect of Use Class 5 are either to use very high loadings of preservatives, in low Natural Durability timbers that are quite easy to treat (such as Pine of course), or to use timbers which have a very high Natural Durability – that is, only those rated as Class 1 to BS EN 350. So your choices will be restricted to such softwoods as European redwood, or 'Douglas Fir' (*Pseudotsuga menziesii*) – which of course, is not actually a true Fir!: and both of those *must* be treated with very high loadings of preservative to a Use Class 5 specification. Or instead, you could specify a heavy constructional Hardwood such as Opepe (*Nauclea diderrichii*) or Greenheart (*Ocotea rodiaei*): either of which would give you at least 50 years as a minimum service life in the sea, without any chemical preservatives being needed.

And once again, please remember that just asking for 'treated timber' is no help to anyone: you *must* specify the appropriate wood species to be 'treated to Use Class 5'. But also note this point, that if you decide to use a Hardwood with a very high Natural Durability (and these timbers actually *are* all Hardwoods, as used for jobs in this particularly demanding Use Class), then you must remember to also specify that any and all of the sapwood which may be present on any timber member must be removed. That's because the sapwood – of any species – has no great Natural Durability at all (as I mentioned earlier on).

So now it's time for the recap on what I've explained in this section.

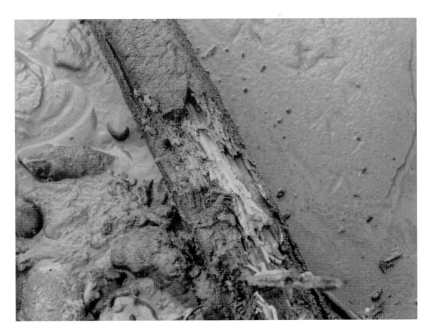

Figure 4.12   Toredo (shipworm) attack

4.11 Chapter summary

In this chapter, I told you that you will need to consider at least three fundamental things about timber when using it for any purpose whatever. These three things are: its behaviour under different conditions of moisture, its Natural Durability rating, and its ease or otherwise of accepting preservative treatments.

You will then need to balance these individual properties against the different conditions where you are intending to use the timber – indoors or out. And to help you to come to a sensible decision about specification and relative costs, I introduced you to two European Standards that you should find very useful: BS EN 335 and BS EN 350.

I also said that it is an odd (and not always helpful) coincidence that each of these Standards list five Classes within it. So remember – it is BS EN 335 which details the five so-called Use Classes, and through these, describes the particular – and increasingly deleterious – hazards to which wood in service may be exposed. And it is BS EN 350 which details the Natural Durability and Treatability ratings of a large (but by no means exhaustive) number of timbers that are in use throughout Europe.

Having weighed up all of these factors – and of course, having also checked the genuine 'risk' level against the particular theoretical 'hazard' – you should then be in a much better position to decide upon an appropriate choice of wood species, either with or without the need for any particular degree of preservative treatment. Remembering of course, that in some of the lower Use Classes, such treatment may be purely an 'insurance' against future problems; whereas in the higher Use Classes, suitable wood preservatives may be the only way of achieving those performance levels with the cheaper, more readily-available commercial Softwoods.

And the final point that I want to drive home with you is that in future, please, please *do not* simply refer to 'treated' timber in a specification or order. You should *always* specify that treatment must be to a relevant Use Class in BS EN 335; and then any timber component must be treated to withstand that Use Class for a minimum desired service life of 15, 30 or (if applicable and possible) 60 years. I hope you will do that properly from now on.

5 The Quality of Timber: Grading for Appearance

I've now brought you to the point where you ought to be able to evaluate any particular situation where you plan to use timber components or wood-based products; and so from that more informed position, you should now be able to make even more sensible and informed decisions, in respect of four important factors. But I make no apologies for very briefly repeating these items at the start of this chapter: partly because they are indeed important in any event; and partly because you may have turned straight to this chapter in order to find out more about grading timber. But you'll have to put up with this little recap first because, as I hope to show, you need to sort out these four factors first, before you go on to take the next step with your specification. So with that in mind, these are the things that you need to decide upon first.

(1) What is the Use Class that applies? And if there is a high or low level of hazard involved, whereby the wood may then be at risk of 'biological degrade' (that's just a Wood Scientist's way of saying 'rot' or some other kind of natural attack, such as Shipworm in marine applications): then what is the probable level of that risk – is it hardly at all, or severe? – and then how can you best plan to minimise it?

(2) Having sorted out the first step and thus decided whether or not there is a need for some sort of resistance to some biological degrade in what you are using the timber for, then you must decide if you want to – or perhaps you need to – specify a type of timber with an appropriate degree of Natural Durability to it. Then any risk of decay (or maybe, marine organism attack) will either be effectively eliminated, or at least, greatly reduced. And there may well be some aesthetic considerations here, as well as physical or biological ones: because the timber component or structure may be a candidate for showing off an attractive wood species, with a beautiful colour or a particularly nice figure: so the reasons for your choice of wood species may then be extended beyond just the need for it to have a long 'service life'.

Wood in Construction: How to Avoid Costly Mistakes, First Edition. Jim Coulson.
© 2012 John Wiley & Sons, Ltd. Published 2012 by John Wiley & Sons, Ltd.

(3)   This step is – in a way – a sort of alternative to the previous step: but it is more than just an alternative, too. Instead of using a durable timber – either a Softwood or a Hardwood – you may have decided, for reasons of cost or some other practical considerations (such as feasibility of access for any future maintenance) to go for some preservatively treated timber instead: and this will usually be a species of Softwood. The extra factor to consider here is the 'treatability' aspect of the timber: in other words, can you pump enough preservative into it, to do the job you need it to do; and for as long as the specified design life needs, for it to survive intact?

(4)   Finally in this 'mini reminder': what must you decide, in respect of the moisture content in your timber? Does the particular Use Class that you've decided upon, require you to dry the timber down to an especially low level, such as below 10 % mc? Or will simply using air-dried timber be good enough – assuming you've given some consideration to the conditions likely to be prevailing at the time the wood is delivered, relative to its final 'in service' EMC. Or maybe on the contrary, perhaps the wood needs to be allowed to *gain* some moisture for its intended use. And I'm sure you'll appreciate that any timber intended to end up fully immersed in water – such as Opepe used for lock gates – doesn't actually need to be fully air-dried before you install it!

OK – so now you're pretty straight on what you need to do, in order to get your timber to behave itself in service for a decent length of time. Then what else could there possibly be, that you need to decide on? Well, how about the precise quality of the timber that you want to use on the job? Does it matter to you – or your client – what it looks like, or how strong it is? I should hope so!

5.1   The need for grading

Let me now give you a couple of examples of why you should be considering the quality or the grade of the timber that you've chosen to use.

European redwood, preservatively treated to Use Class 3, will be fine for use as external joinery, as I told you in the last chapter. But even though the wood may last you a reasonably good length of time, would you (or your client) be happy to have numerous clusters of large knots all along the length of each of your window frame and cill members? And even if you would be prepared to accept the presence of *some* knots, how big would you expect those knots to be? – just a few quite small ones, maybe? Would you also allow some splits to be visible on the exposed surfaces? – and how big could those be? And would you allow the timber supplier or joinery manufacturer to include occasional pith streaks along some exposed surfaces, as well?

Let me try another example. Say you're taking delivery of some trussed rafters, made from European whitewood (i.e. Spruce), for a domestic roof. Would you then allow some knots or some splits in those? And if so, where, and how big? Would you also permit some pinworm holes? And if so, how many and over what length of the timber component would you allow these possible 'defects' to be present in?

I'm deliberately being a bit provocative here: but those not-over-fanciful examples given above are actually intended to highlight the thinking behind the European Standards on timber – some of which I want to touch on in this and the following chapters. I want to get to the bottom of what is really meant when different people in the Construction Industry use those somewhat misunderstood words 'Quality' and 'Grade'.

5.2 'Quality' or 'grade'?

What I'm dealing with here is the fact that every single piece of timber varies, in terms of exactly what you will get out of a log when it is being processed into various lengths and cross-sectional dimensions, in a sawmill. And of course, there will always be a certain element of conflict here, between the saw miller and his intended customers – who are generally the timber importers. And there is also a conflict between the Timber Trade and their own customers – who are mostly the construction industry (by which term I also include Joiners and Shop Fitters, not just Carpenters or Builders).

It is a 'given' that saw millers would very much prefer to sell everything they can produce, at the highest price; whereas their customers – or their customers' customers – would prefer to buy every stick of timber from the mill or the wood yard at the lowest possible price. Common sense tells you that nobody is going to pay the same amount of money for a rough-sawn, splintery, bent and knotty lump of wood as they are for a smooth, planed, straight and clear piece of the same stuff. ('Clear', by the way, is the term used to describe more or less knot-free or defect-free timber: as I will discuss in a bit more detail later).

Since nobody wants to pay too much for their wood, but the saw miller wants to get as much money as possible for his production, the logical thing for the saw miller to do is to sort the outturn from the mill's production into several different qualities or grades; and then to price these different grades accordingly, based upon various (and often quite different) factors. These factors will be things like the scarcity of a particular species of wood, its highly decorative surface appearance; and perhaps some non-appearance-related factor, such as its availability in very large cross-sectional sizes, or maybe its ability to support a given load as a structural member (but see the next chapter for this particular aspect of grading).

So how exactly do 'quality' and 'grade' differ anyway? Or do they indeed differ at all? In many parts of the Timber Trade, you may hear these two terms being used interchangeably: though I personally would prefer to be a little more exact about it (as you might have guessed by now).

5.3 Quality

To my way of thinking, the 'quality' of anything really means its 'fitness for purpose'. For example, a fence post can get away with a lot more in the way of knots, splits and discolouration than a widow frame can. So it should stand to reason that the overall appearance of a batch of fence post timbers will be considerably lower than will the appearance of a pack of joinery timber, in terms of the visual impact of the wood itself. And yet each of these two batches of timber will still manage to be the right 'quality' for the job in hand.

How then, does the saw miller decide upon all these different 'qualities'? He (or she) will do this by sorting out or selecting the timber at a particular stage during the production process and then keeping those various selections separate – and yet consistent – within a range of identifiable parameters. Such selection parameters are usually published these days (although they may not be readily available to the outside world!); and all of us in the Timber Trade and its associated disciplines, refer to these published parameters as 'Grading Rules'.

5.4 Grade

That very process of 'sorting out' the timber into the various 'use qualities' (so to speak), based on such things as appearance, is of course what is understood as the whole idea of 'grading' (see Figure 5.1). (I might add, out of interest, that the German word for 'grading' is 'Sortierung' – so you can see the linguistic link

Figure 5.1   The grading line in a modern sawmill

there.). So think of the difference in those two terms in this way: 'quality' is more concerned with the particular end-use suitability of any timber; whereas 'grade' is the method by which the timber was actually selected as being potentially suitable for the job. And from this, you can see that the grade really acts as a sort of description of what you might reasonably expect the timber to look like, in a very general sort of way: somewhere between pretty and or ugly (or just pretty ugly?).

If you will now try to hold in your mind my explanation of the subtle difference between quality and grade, I promise you that it will help you to better grasp the elements of what comes next.

5.5 The different types of grading

As I've just discussed above, there is almost always a need for doing some sort of selection or grading of timber, in order either to sell it in the most market-efficient way, or for the customer to be able to differentiate it on price. And to do this, there are three basic methods of grading that are employed. (Well there are really two main methods: but the first method is usually sub-divided into two variants, primarily according to timber type. I will deal with those two variants of the first method here; whilst leaving the second main method until the next chapter.)

5.6 Appearance grading

As this term would perhaps obviously imply, the sawn timber is sorted out (i.e. 'graded') on the basis of how 'good' or how 'bad' it looks to a potential customer. I should immediately point out that these 'appearance' grades are not generally sold as being fit for any particular purpose – such as joinery or fencing. The final decision concerning the end-use of any timber is normally left in the hands of the final purchaser. But the various grades which the different Scandinavian, European or North American sawmills can offer have, over time, become more or less accepted as being fairly appropriate for a reasonably specific range of end-uses.

As I was just telling you, there are two main variants, or sub-divisions, of this 'appearance' type of grading: and they are referred to as either the 'defect' system or the 'cutting' system. The defect system is generally used with Softwoods, whilst the cutting system is normally used to separate out attractive and decorative Hardwoods; although there are also a few North American grades of joinery Softwoods which are selected, and offered on the market, by means of the cutting system.

5.7 Appearance grading: based on defects

The word 'defect' may strike you at first as a bit of an odd term to use. After all, who would knowingly want to buy a 'defective' product? But in the particular context of timber – which is a natural and highly variable material – this term

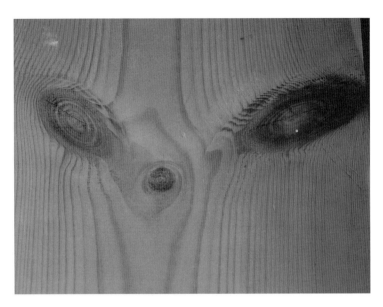

Figure 5.2   A typical knot cluster

really means anything inherent in the wood that can affect its performance, or its acceptability, in some way. These 'defects' may be things which were already there naturally – due to the particular way in which the tree grew – or they may be things which were introduced as a consequence of the sawmilling operation, or some other related production process, such as drying.

Natural defects can therefore include such things as knots (from the tree's branches), wild grain, ingrown bark or resin pockets (see Figures 5.2 and 5.3). Defects that arise from processing can include wane (that is, one or more corners or edges missing from a board: because we insist on converting round-shaped trees into square-edged timber); drying cracks or splits; and distortion (see Figures 5.4, 5.5 and 5.6). (This last defect usually happens as a result of the wood drying down too rapidly; but it also may be made worse by a lack of straight grain or other natural growth defects in the timber, as well.) There are numerous permutations (which are generally related to their size, length or visual impact) of these so-called 'permissible defects', some or all of which are allowed to a greater or lesser extent, within the myriad selection of the different Commercial Grades that are to be found amongst the huge volumes of timber that we use and import into the UK each year. (8–10 million cubic metres per year of Softwood alone, during each year of the first decade of the new Millennium!)

It should come as no surprise that different grading systems in different parts of the world actually apply different rules to the same defects: and thus define to what extent any specific defect is allowed, in any specific grade. But in general terms, all of the grades work to the same basic principle: that the highest

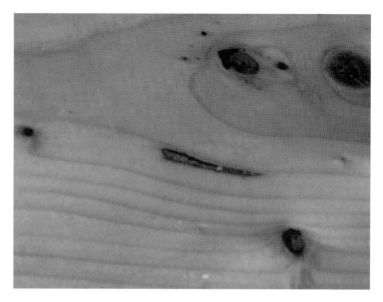

Figure 5.3   A resin pocket in spruce

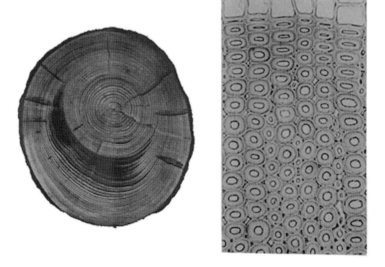

Figure 5.4   Compression wood – in the log and in microscopic section

(best) appearance grades will allow smaller amounts of everything within them; and vice versa for the lowest (worst) appearance grades, which will of course allow much greater amounts of everything to appear within them.

I don't propose to spend any great length of time analysing the actual physical process of grading, as it is practised in sawmills from Latvia to Louisiana, or

Figure 5.5   Many boards showing wane on one or both edges

Figure 5.6   Distorted timber

wherever – because the exact methodology behind producing the different qualities needn't concern you over-much in a book like this, whose main aim is to try and help you to specify and use timber in a better way. But in order that I can help you to improve your knowledge about what you are hoping to do with wood, it is definitely worth spending a little while explaining some of the – often quite odd and confusing – terminology that is used to denote the various timber grades that are produced in Scandinavia, Europe and North America. I will then attempt to compare how (or even *if*) they might match up to one another.

5.8   Scandinavian grades

Sweden and Finland have been sorting timber into many different appearance grades (or qualities) for well over 100 years; and in that time, they have evolved a system of nomenclature that can, at times perhaps, seem a bit antiquated to us now from our current lofty perspective here in the 21st century. Because of their very long history and linguistic evolution, these rather odd terms – which probably still made perfect sense in 1911 – may cause a bit of confusion in 2011 and beyond.

Very originally – that is, a few hundred years ago, when there were still very large, virgin forests to go at – the Scandinavians were able to produce good quantities of high quality 'clear' (i.e. virtually defect-free) timber. (I must quickly interject that this is not to say that Scandinavia doesn't have large forests any more: far from it, in fact. But today's forests are, by and large, managed rather than completely virgin and natural; and they are frequently second-growth forests, having been harvested originally in the mid to late 19th century and then replanted or re-seeded – and thus very happily re-grown for us today.) But you may now understand that Scandinavia was already exporting wood to other parts of Europe, as far back as at least the 16th century: and over the past few hundred years, they evolved their present system of timber grades and qualities.

The Scandinavians were, in those long-ago times, quite prepared to offer entire parcels of the very highest grade of timber to their customers: and especially for export. But gradually, over the intervening years – and well before the Second World War, in fact – the available quantities of really large diameter trees have become more and more limited. And in addition to this, large areas of newer, second-generation forests have been extensively planted throughout much of Scandinavia and Northern Europe: some forests even having already been re-planted for a second time! And during this time, new felling regimes – designed to 'rotate' the different forest areas and so give them time to fully recover – have been instigated. All of these factors have resulted in a considerable reduction in the availability of large volumes of the very highest quality, 'clear' timber. As a result of these gradual changes, nobody today in Scandinavia or Europe ever offers their customers entire packs of the 'very best' quality of wood any more.

5.9   Unsorted, fifths and sixths

As I began to explain earlier, the main commercial 'appearance' grades of timber from most of Northern Europe were originally available to their customers in each of six distinct 'quality' divisions. These were always (and even to this day, they still are) designated in Roman Numerals: going from I ('Firsts' – which denotes the best-looking) all the way down to VI ('Sixths' – which is considered to be the worst-looking).

But – bearing in mind what I was just been saying – there is no longer the possibility of the Timber Trade being able to buy an entire pack of 'Firsts' from any Scandinavian or indeed, any European producer. The best available quality of timber that is now on offer from any Nordic sawmill will be a mixed assortment, consisting of the top four grades, all put together within every pack: and this is universally described under the highly confusing label of 'Unsorted'.

Now, as a relative layman, you could be entirely forgiven for jumping to the obvious conclusion that a quality of timber that is called 'Unsorted' had never been … er, um … sorted. Or graded, at any stage of its production. But you'd be wrong. This apparently daft term, which has unfortunately been perpetuated by the Timber Trade for over 100 years – and which seems to imply that no 'sorting' or 'grading' has ever been done to the timber – in fact means exactly the opposite! So, when it comes to Unsorted timber, what you are really being offered are the first four of the very 'best' appearance qualities of timber, all lumped together as the one, current, 'highest available' quality – and its strange name simply means that it has never been subsequently sorted or separated into its four distinct component qualities (which are I, II, III and IV). It has instead been given this incredibly confusing name … indeed, I sometimes feel that the Timber Trade is deliberately trying to confuse us all!

So now, if the top four qualities are being sold off together as something called 'Unsorted', you can perhaps begin to understand why the next available quality down the scale is a grade described and sold as 'Fifths'. This is (somewhat obviously now, when you have it described thus!) because the first four grades have been 'swallowed up' in the Unsorted title and so the next one that is sold *on its own* is – quite naturally – V (Fifths). And then this next-lowest grade of Fifths is followed (a lot more logically now, you might agree) by the very lowest Export appearance quality, which is known as 'Sixths' (and written VI). These last two numerical terms seem to emerge quite naturally from the basic range of *six* qualities; but they can be somewhat puzzling on their own, when taken out of context, and when they appear to follow on from something apparently quite unrelated, that is called Unsorted.

I hope that's now clear? Well, I hope so – because just when you thought it was all making sense at last, here come the Russians.

5.10   Russian softwood qualities

Having just finished telling you that the Scandinavians have six named qualities (of which only three exist in reality, of course), I will now tell you about the exception to that rule. The Russians (bless 'em) decided that they would use only *five* basic qualities to describe their timber: and so *their* 'Unsorted' consists of the first *three* qualities mixed up together. Then out of that arrangement comes *Fourths* as the next lowest, and finally *Fifths* at the bottom of the appearance quality heap: so there is NO Russian 'Sixths' (or at least, not as far as their Export Market is concerned: but let's not go there, please!).

Now let me try to be clear about this. Scandinavia divides its available timber qualities – based entirely on their visual appearance – into six basic selections: but Russia takes essentially the same wood, and yet it creates only five basic selections. Scandinavian Unsorted consists of grades I–IV inclusive; whereas Russian Unsorted consists of their grades I–III inclusive. The penultimate low grade in Scandinavia is Fifths, whereas in Russia it is Fourths; and the very bottom appearance grade in Scandinavia is Sixths, whereas in Russia it is Fifths (see Table 5.1). Got that now? Good!

You may well say, 'Well, what's the point of all that nonsense about Grades?' Fair enough, if you don't have to buy them every day, I suppose. But the point is that, if you are ever offered a timber components made from a parcel of Russian Fourths, DO NOT think that the components will be made out of timber from the bottom bit of the 'Unsorted' range: they are not. This timber is of a quite low visual quality, more akin to Scandinavian Fifths. And likewise, should you be offered some Russian Fifths, DO NOT think that they are only a relatively or moderately low grade. They are not: they are only about as good as Scandinavian Sixths (i.e. not very), and so they will most likely be somewhere near the very bottom of the range of acceptable appearance.

Table 5.1   Russian vs Scandinavian timber qualities

Russian		Scandinavian
(Unsorted) {	I II III IV	– Unsorted
(Fourths) IV	V	Fifths
(Fifths) V	VI	Sixths (Utskott)

5.11   European appearance grading

I have referred to Europe in more or less the same terms as Scandinavia: and to a large extent, that is true. But sawmills in mainland Europe these days – especially Germany, Austria and The Czech Republic – will tend merely to pay lip service to the Scandinavian system; and they will often produce their own variants of those three qualities of Unsorted, Fifths and Sixths. But European appearance grades will still, however, follow the essential principle that the top appearance grade is some sort of 'first' quality: then followed by consecutively increasingly higher numbers to denote the lower qualities. Indeed, some sawmills in Germany that I regularly deal with, actually use the symbol 'VI' (Sixths) as their 'reject' grade – thus effectively using that term as a 'catch-all' for any timber that is not considered to be fit for most other uses. The Baltic States (Estonia, Latvia and Lithuania) tend to adopt the Russian system of designating only five theoretical qualities: but of course, with only three grades sold in reality – which are Unsorted, Fourths and Fifths.

These days however, the main tendency with European sawmills is that they will produce specific grades of timber for specific end-uses, and/or for particular export customers: thus they will saw their timber into sizes and qualities that are generally suitable for (say) pallets or scaffold boards, and so on. (Please note that here I have not included 'Strength Graded' timber – I'm leaving that particular subject for the next chapter.)

Now I need to cross the Atlantic, metaphorically speaking, to show you how they do things over there as regards grading their very different Softwoods for different appearance qualities.

5.12   North American softwood appearance grades

The USA and Canada have long cooperated in their grading rules – although they still publish them as separate documents in their respective countries – and there are many grades that exist for their internal markets, in both the East and the West of that Continent. But the two countries have managed to cooperate especially well insofar as their Softwood Export market is concerned: so much so, that they have created a very comprehensive set of rules, which originated on the West Coast in the early years of the 20th century. These grading rules are known, somewhat arbitrarily, as the 'Export R List'. This Rule Book (because that's what it is) is issued by a body called the PLIB: the Pacific Lumber Inspection Bureau. (As an interesting aside: back in the mid-1970s, I once asked a very old and long-retired Canadian grader why this Rule Book was called the 'R' List; and he said it was because its predecessor was known as the 'P' List (see Figure 5.7). But then, nobody knew where *that* name came from – so we're none the wiser!)

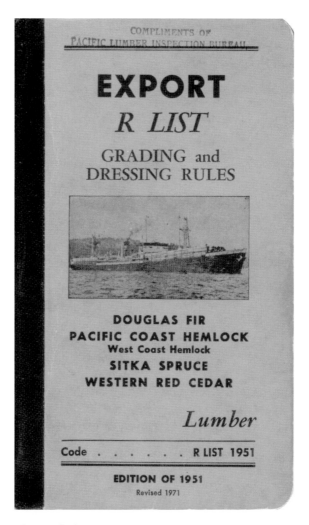

COMPLIMENTS OF
PACIFIC LUMBER INSPECTION BUREAU

EXPORT
R LIST
GRADING and
DRESSING RULES

DOUGLAS FIR
PACIFIC COAST HEMLOCK
West Coast Hemlock
SITKA SPRUCE
WESTERN RED CEDAR

Lumber

Code . . . . R LIST 1951

EDITION OF 1951
Revised 1971

Figure 5.7   Export R list

Anyway – these 'Export R List' rules apply to virtually all of the commercial Softwood species that are grown in North America; and they define a very large number of grades, many of which have never been – and probably never will be – seen in the UK. But I will, in a few moments, describe to you those appearance grades of the North American Softwoods which you might sometimes see on offer in the UK. These particular grades are more especially seen with West Coast wood species such as Western Hemlock (*Tsuga heterophylla*), 'Douglas Fir' (*Pseudotsuga menziesii*), and 'Western Red Cedar' (*Thuja plicata*). (By the way: I hope you've noticed that I've written those last two timber names 'in quotes', since that is the official way of showing that their Trade Names are 'false' and that they do not really belong in the families that their names would appear to place them. In other words, as I've no doubt mentioned elsewhere, 'Douglas Fir' is not a true Fir and 'Western Red Cedar' is not a true Cedar. You may now care to re-visit all that stuff from Chapter 2, when I told you about Timber Names and how confusing they can get: and how the Timber Trade don't help maters, by calling timbers by names that are not strictly correct.)

Many of the wood species included in the PLIB's 'Export R List' – including the three timber species that I just mentioned above – grow as very large trees indeed: many of them being considerably more than two metres in diameter. As a result, these particular wood species are still available in commercial quantities of very high grade and virtually defect-free timber (or 'lumber' as our word 'timber' is called in North America): and which are sold as distinct grades in their own right. And there is a reasonably good range of lower appearance qualities available too, of course.

5.13   Clears, merchantable and commons

The very best appearance quality of timber from North America is called 'Clears'. This means that such a piece is essentially free from all visual defects – although it may actually have a few small knots permitted on its 'back face'. (This is the term used to describe the fourth side of a piece of timber: the other three being the 'best' face – which means across the width of the piece, plus both edges. In other words, these are the three faces of the timber that would be visible if the piece were to be viewed normally in three dimensions, when resting on a flat surface.) You can see from the foregoing that even a so-called 'clear' grade may not be completely blemish-free; since one face can still look a little bit 'imperfect'.

This category of Clears is then further sub-divided into different and slightly lower appearance categories (the names for which mean, in effect, 'near perfect', 'not quite so perfect' and 'just a bit less than perfect'): and in fact – despite what I've just told you about the good availability of very high quality wood from North American sources – there is no longer any amount at all of the 'No 1 Clear' grade being offered as commercially available. So the very best appearance grade – and it is nonetheless a very, very high quality of fine-looking timber – is in reality called 'No 2 Clear and Better': which is usually abbreviated to the easier term of 'No 2 C&B'.

Just below this 'best available' grade is one called 'No 3 Clear': and then sometimes there will be a 'No 4 Clear' grade offered as well: although it is, in my experience, restricted to particular species, such as 'Western Red Cedar' – and even then it is only seen very occasionally in the UK.

After the whole batch of 'Clears' comes a set of slightly less good-looking grades, known collectively as the 'Merchantable' qualities. The best of these is called 'Select Merchantable' (normally abbreviated to 'Select Merch'); and this is then followed by 'Number 1 Merchantable' and then 'Number 2 Merchantable' (naturally, these latter grades are abbreviated to 'No 1 Merch' and 'No 2 Merch' respectively). As their collective name of 'Merchantable' might imply, they are not meant to be fully defect-free; but they are still supposed to be very saleable and highly usable; and thus fit for making some products or components that can then be sold on to somebody. All of these Merchantable grades – being lower than the Clears – will contain some knots of varying shapes and sizes; and they will also contain greater or lesser amounts of other visual defects, of any of the types that I described earlier on in this chapter.

The final PLIB appearance grade that is occasionally seen in the UK (although it is pretty rare to see nowadays), is 'Number 3 Commons': and this is not usually abbreviated in any way. It is – as its name perhaps implies – a pretty rough-looking grade, within which all sorts of defects are allowed: including even wormholes and pocket rot. (Yes, I did say rot!)

5.14   A comparison of Scandinavian grades and North American grades

I promised a little earlier to try to compare the main appearance grades from Scandinavia with those from North America. Bearing in mind that there are no effectively 'clear' qualities on offer from any European sources, you will not be surprised to learn that there is no exact match between the two sets of grading rules for Northern hemisphere Softwoods. The table that I have given you here is merely my attempt at showing you where the nearest similarities may be: it is not to be taken as any kind of 'official' scale of comparison (see Table 5.2).

Table 5.2   Comparison of Scandinavian and North American softwood qualities

Scandinavian grade	North American grade
No match	No. 2 Clear and Better
Best Unsorted	No. 3 Clear
Good Unsorted	Select Merchantable
Lower quality Unsorted	No. 1 Merchantable
Fifths	No. 2 Merchantable
Sixths	No. 3 Common

5.15   Appearance grading: based on 'cuttings'

I said earlier that 'appearance' grades can take one of two essentially different formats. The first kind – as I have been outlining in some detail above – is based entirely upon how pretty or how ugly the wood is (and some of it can indeed be pretty ugly, as I hinted before!). The philosophy of the so-called 'defect method' is that entire packs and indeed even the individual pieces within those packs are always sold in their finished, sawn dimensions 'as seen': and thus without any great consideration as to what that particular grade or quality of timber may ultimately be used for.

However, the second type of 'appearance' grading – known almost universally as the 'cutting' or more usually, the 'clear cutting' system – has evolved specifically to take account of the expected range of end-uses of the timber. This is catered for by allowing the presence of limited defects in certain locations, and/or in particular quantities or combinations, in a limited number of places along the length of each piece within a parcel of that grade. This can be done because each piece is individually graded, on the understanding that such defects will then later on be 'cut out' of the individual member: then leaving a percentage of 'clear' timber to be eventually made into something, by the final purchaser. And this process – as you will hopefully now see – is the origin of the term 'clear cutting'. It is quite complex to grasp – and indeed, to perform – since it relates to both the *number* of cuts allowed and also to the *position* of those cuts, in each and every piece of a specified grade.

Although there are just a few Softwood grades based on the cutting system (and which may be found in the PLIB 'Export R List'), the vast majority of instances where this 'clear cutting' method is used will be with Hardwoods. And there are two principal sets of 'clear cutting' rules used in the world for grading Hardwoods.

One set of rules relates to Temperate Hardwoods – and they originated in the USA – and the other set of rules relates to Tropical Hardwoods – and they originated in Malaya, at the time when it was part of the British Empire (but which is now of course the independent country of Malaysia). And indeed, these latter rules were very originally known as the 'Empire Grading Rules for Tropical Hardwoods'; which then became the 'Malayan Rules', prior to that country's independence.

These two very different sets of rules these days are known respectively as the National Hardwood Lumber Association (of the USA) – or 'NHLA' Rules – covering Temperate Hardwoods – and the Malaysian Grading Rules – or 'MGR' – covering Tropical Hardwoods. And although they originated, as I have said, in those specific parts of the world, these two sets of rules are now used pretty well universally – although sometimes in a modified form – for all appearance-graded Hardwoods, traded the world over.

5.16 The NHLA grades

The latest copy of the NHLA Rules that I have on my bookshelf is dated 2003 (see Figure 5.8). These rules don't change all that much, but they are revisited every decade or so, just to see whether they require any updating at all. Unsurprisingly, not all of the possible (or even available) NHLA grades get as far as the shores of the UK; although the better qualities usually seem to. However, as with other grades you've learned about so far, even the very best

Figure 5.8   The NHLA Grading Rulebook (2003 edition)

quality that is available under these long-established Rules is not a fully 'clear' grade of Temperate Hardwood: it still permits a few defects (mostly things like small knots or bark pockets), provided that they are indeed fairly small and can be easily cut out, without chopping up the graded member into such tiny pieces that it cannot be realistically used for any meaningful job.

As I said earlier, the cutting method relies on the fact that the defects are only there in the first place because they are *intended* to be cut out by the final user; and because of this, all of the specifically-allowed defects are restricted: both in their size and location along the board. But as well as this, the total number of cuttings allowed in any board, to remove those defects, is also limited. And not only that; there is a third factor at play here as well. In addition to the requirement that any piece of a certain grade must yield a minimum usable percentage of clear wood, *each individual piece* must also eventually yield cut pieces of timber which meet minimum specified lengths and widths, after all of the defects have been cut out. A look at the most commonly-found grades as exported to the UK may help to clarify what I've just been saying.

5.17   FAS, selects and commons

The very best quality of Temperate Hardwood available under the NHLA Rules is called 'First and Seconds' – but this is always abbreviated to 'FAS' and its title is never quoted in full. You've probably noticed that, as with the other grades that I discussed earlier, there is no individual category of a 'First' quality available any more (if indeed there ever was, since the origins of the NHLA Rules are shrouded in some considerable vagueness). Despite the lack of an individual 'First', the FAS grade is still of a remarkably high order of magnitude when measured on the 'clear wood' scale. It restricts the number and size of any defects to those which, when cut out, must still yield a minimum of just over 83% – or 10/12ths – of perfectly clear timber. (And in fact, some boards may even have 100% clear, completely defect-free timber in them.) And that's also on *both sides* of the piece! But in addition to the requirement for a minimum percentage of final 'clear' yield, there is also a requirement that the individual pieces left after cutting must be a minimum of at least seven feet long by three inches wide, or at least five feet long by four inches wide. (That's long enough to make door stiles out of, for example.)

You won't have failed to spot that the final minimum usable dimensions under the NHLA Rules are all given in Imperial Measure: and that's because our cousins across the Pond decided to give up on going Metric, some time back in the 1980s – and so they can still delight in using feet and inches, when they sell 'lumber' (which, remember, is their word for timber) to the UK Timber Trade.

(I will do no more than mention here that American Hardwood Measure is also a bit of a Black Art; dealing as it does in something called 'board feet', and with thicknesses stated in such ways as 'six quarters' (??). But please, let's not go there, in this simple little handbook!) Indeed, the Hardwood Trade as a whole tends to delight in feet and inches, whilst the rest of the Timber Trade uses Metric: with greater or lesser enthusiasm, it has to be said …

5.18   Selects

Now to continue with the NHLA Grades: the next quality down from FAS is known as 'Selects'. (And please don't confuse this grade with the 'Select Merchantable' grade of Softwoods from the PLIB 'R' List – though I know that would be easy to do!) This particular NHLA grade that is called Selects is very similar to FAS, but the

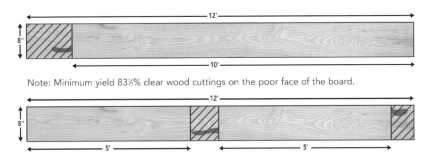

Note: Minimum yield 83⅓% clear wood cuttings on the poor face of the board.

Figure 5.9 FAS grade illustration (source AHEC)

minimum sizes after any cuttings are different – and just slightly shorter – being a minimum of six foot lengths by four inch widths (see Figure 5.9).

The next 'official' NHLA grade down the list is called No.1 Common. This grade includes boards that are a minimum of four feet long and three inches wide; and it will yield clear face cuttings of just over 66% – or 8/12ths – that is, up to, but not including, the minimum percentage requirement for FAS: and the smallest clear cuttings allowed from this grade are three feet long by three inches wide, and two feet long by four inches wide. For this reason, it is prized as a grade suitable for producing fine cabinet work (and indeed, it is sometimes known as 'Cabinet Grade'). The total number of permitted clear cuttings permitted to achieve this grade will be determined by the final size of the boards produced.

5.19 'F1F'

Below the level of FAS and Selects, there is a sort of 'intermediate' grade which is actually a combination of FAS and No 1 Common: and it is known as 'FAS 1 Face' – or more usually, just 'F1F'. With this grade – as its name would suggest – a board's best face must be a full FAS, whilst its back face must be not less than a No. 1 Common in appearance and cuttings.

5.20 Prime and Comsel grades

These are two Export grades that are not in the NHLA Rule Book, as you will find out if you ever try to read it in any detail. The first of these – the so-called 'Prime' grade – is an 'unofficial' grade which evolved from the FAS grade. Boards that are considered to be 'Prime' must be square-edged or virtually free from wane. In its yield, it is more or less equivalent to a sort of 'Select and better', with a very good overall surface appearance being the major factor. The minimum allowable sizes of any boards will vary, depending on the species, the region of the USA where it is sold from, and the producer who supplies it: since this is not a 'proper' grade and so it has no fully defined limits within the NHLA Rules.

'Comsel' grade is another 'unofficial' grade that has evolved for the export market: but this time, it has developed from a combination of 'No. 1 Common' and 'Selects' – hence its fairly obvious composite name! Its minimum clear yield should be at least as good as No. 1 Common, or preferably slightly better, and with good surface appearance as its main criterion. As with Prime, the minimum board sizes will vary, depending on the species, region and supplier; since this also is not a 'proper' grade, and again it is therefore not defined under the NHLA Rules.

So much for the grading of Temperate Hardwoods: now for the Tropical ones.

5.21   Malaysian grades

The Malaysian Grading Rules – as I mentioned earlier in this chapter – have been in existence since the British Empire days; and they have been revised at different intervals, ever since Malaysian independence. The latest revision was published only as recently as 2009: and these Rules define a number of grades for Tropical Hardwoods; although as ever, only a few of them seem to make it into the UK. They are – exactly like the NHLA Grades – based on the concept of 'clear cuttings': with a minimum percentage yield and certain minimum lengths required for each grade, after cutting out the defects.

They are also perhaps slightly easier to grade, and somewhat easier to obtain from the log, than are the NHLA Temperate Hardwood grades: that's because Tropical Hardwood trees have very few low-level branches and they mostly produce larger diameter logs than the Temperate Hardwoods do. Therefore they produce fewer knots and have a much smaller proportion of sapwood within the trunk. But, on the down side, the Tropical Hardwoods are a lot more likely to suffer from worm attack and sap stain, due to the very warm and humid tropical climate. Oh, and there is one more recent change to the MGR book: all Malaysian timber graded to the MGR will now show a stamp mark bearing the initials 'MTIB' – the Malaysian Timber Industry Board. Other tropical species from other countries may use the MGR system and the grade titles for their grades, but they will not, of course, carry the MTIB stamp.

There are at least six Grades available under the MG Rules, but only the first three – with a variant that I will then explain – are likely to be seen in the UK.

5.22   Prime, select and standard

The top grade under the MG Rules is called 'Prime' – and it must result, after the allowable cutting has been done, in boards of six inches and wider and six feet and longer. You will by now be quite used to the fact that a single word can have many meanings within the Timber Trade: so you won't be surprised that 'Prime' under the MG Rules is not at all the same thing as the 'Prime' that you just encountered in connection with the NHLA Grades!

The same goes for 'Select': another of those seemingly universal words to do with timber quality, but which in the context of the MG Rules has a particular – but of course, different – meaning. It is only just below Prime in appearance quality and must result, after cutting, in boards of five inches and wider and six feet and longer.

'Standard' Grade is the third-best of the MGR grades: and it allows larger knots than the two grades above, but it is otherwise much the same in respect of other allowable defects. However, it results in very many smaller pieces after cutting; yielding boards four inches and wider and six feet and longer.

5.23 'PHND', 'BHND' or 'sound'

This is the variant that I mentioned earlier: and in fact, all of these terms refer to one and the same thing. The first two sets of initials stand for 'Pin Hole No Defect' and 'Borer Hole No Defect': which means effectively that wormholes (but not live infestation) will be allowed in the cuttings; and then the presence of such holes may not be used as a reason to reject the timber. 'Sound' Grade will be wormy, but otherwise similar in appearance to the first three grades above; and its board sizes will be comparable, depending upon the exact specification stated by the producer.

5.24 Rules are made to be bent! (within reason)

One final but quite important point needs to be made, as far as all of the so-called 'appearance grades' are concerned. That is, that they are intended to more or less regulate the type, size and number of both natural and process-induced defects which are allowed to be present in any length of graded timber of a specified quality. Therefore, a customer is not just buying 'wood' when he buys a pack – or even just one length – of (say) Unsorted Swedish Redwood or FAS Grade Black Walnut. He is actually buying something with a fairly defined level of visual appearance – and an appearance which will indeed be *expected* by those who are then buying it or using it. That is, if they have sufficient experience of what they think they are actually buying or using! But although the different sets of rules will define the allowable defects in different ways, and within reasonably accurate limits, they are still flexible enough to allow for variations in the final board appearance; and this may be as a result of differing interpretations of the rules, or different qualities of logs available to the sawmills, or a number of other factors.

5.25 Shipper's usual

There will (of course!) be variations in the precise appearance of the timber from one piece to another: after all, an Unsorted quality can include anything from a First to Fourth within its grade boundaries. And overall, there will still be a minimum acceptable appearance of the timber, which buyers will be familiar with receiving, from particular Shippers (i.e. producers) that they regularly deal with. And this 'expected' level of reasonably consistent quality from any one producer is known to the Timber Trade as 'Shipper's Usual'. It is, in effect, some particular producer's own interpretation of whichever set of Rules he is working to, based on the timber that he is able to cut and grade, from the logs that he is able to buy.

Because of this 'natural variability' within the source material, a Northern Finnish Unsorted quality of (say) redwood, is likely to be very much better-looking than a South Swedish Unsorted quality of the same wood species. (In fact, even a far-Northern Fifths quality from any Scandinavian country, is very likely to be a lot better-looking than a South Swedish or a Polish Unsorted quality – simply because of the quality of the saw logs that are available to the sawmill in each of those locations.)

The lesson to be learned by the timber buyer from all of this is: *don't just buy on price*! That's because the final visual quality of the sawn timber can vary one heck of a lot – even when comparing grades that are ostensibly supposed to be the same thing. So you' be much better advised to inspect a trial batch of any new timber ('new' that is, in the sense of either/or the species and the Shipper) before you commit to buying lots of it, if that is at all practicable. Then, and only then, will you *really* know you've got a fair deal for what you're paying for, as a buyer of timber to a particular description. But what if you're a specifier of components for joinery or shop fitting, who doesn't want to get embroiled in the niceties of the descriptions and qualities of timber that the importers and merchants deal with? What should you ask for if you want to make sure that your wooden windows or door frames look good?

5.26   BS EN 942: the quality of timber in joinery

The answer for the specifier of timber used in joinery is to go by the qualities of timber given in this helpful European Standard: which has now replaced our 'old' BS 1186 descriptions of the qualities of timber used in joinery. No longer must you speak of 'Class 1 Joinery' and so on … now you need to be asking for 'J' Classes of wood.

5.27   J classes

It should be pretty obvious that the 'J' Classes have be so named, because they are deemed as being suitable for use in joinery of differing appearance qualities. These Classes cover both Softwoods and Hardwoods – and indeed, wood-based sheet materials as well: but a cursory glance at the allowable defects within each Class will soon make you realise that they were really written with Softwoods in mind (see Table 5.3).

**Table 5.3   Joinery classes as given in BS EN 942: 2007
Example of Timber Classes, showing maximum permissible knot diameters/
face percentages**

	J2	J5	J10	J20	J30	J40	J50
Max knot diameter (mm):	2	5	10	20	30	40	50
Max % of knots on exposed face:		10	20	30	30	30	40	50

(NOTE: The smaller of these two limits always applies: thus in Class J2 to BS EN 942: 2007 for example, a 15 mm wide component may only have a 1.5 mm knot on an exposed face, but a 50 mm wide component may only have a 2 mm wide knot on an exposed face.)

The 'best' Class is J2, which is effectively 'clear' – in that it only permits knots of 2 mm or smaller in diameter ('pin knots' as they are known). This has replaced the old 'Class CSH' of BS 1186, which stood for 'Clear Softwoods and Hardwoods'. The next Class is J10, which allows 10 mm maximum knots *or 10%* of any exposed face: and so on, up to J50 – which of course allows 50 mm maximum size knots, or knots covering up to 50 % of any exposed face; which ever is the lesser.

5.28   'Exposed face'

A word of caution here – for both the manufacturer and the specifier of timber used for joinery. The sizes of defects which are allowed in the 'J' Classes relate – as I have said – to the 'exposed face'; but this is not the actual surface of any element which is visible *in the finished item*, and so these defects relate to the *full dimension* of the component. I'd better clarify that a little.

   Take a bottom window member, which incorporates a cill, and that has been machined out of (say) 63 mm × 150 mm redwood. It will end up as a profiled component with dimensions approximately 58 mm thick and 145 mm wide at its greatest extremities (and of course much smaller dimensions in certain places; such as the front edge of the cill). And let's say that the cill – which is the part you will see when the window has been fully assembled – projects 45 mm out from the front of the window frame. Now imagine a knot somewhere in the exposed cill part of this moulding: how large can it be, in a J50 quality of timber? 22.5 mm? (That is, 50 % of the 45 mm cill which you will see in the completed window assembly.) No! The cill member can actually contain one or more knots of 50 mm maximum diameter – for the simple reason that you must measure the defect against the *full* dimension of the component – in this case, the 145 mm finished size – and so the rules permit 50 % of 145 mm or a 50 mm knot, whichever is the lesser. So it is possible to have a knot more or less the *full width* of the exposed cill in J50, because that cill is part of a larger member and the rules apply to the whole piece, not just the exposed bit of it.

Of course, with Hardwoods, the knots don't really enter into the picture (literally!); and your final appearance grade will be influenced by other visible things; such as surface cracks (collectively called 'fissures', remember), wane, resin or bark pockets, and so on.

The thing to remember about the 'J' Classes of BS EN 942 is that the criteria for visual defects apply *only* to the finished joinery item and not to the stock of wood that it was made from. At first reading, that sounds a bit daft: but what I mean to say is that it is *impossible* – and I mean that word literally – to ask for, or to grade out, a pack of 'Class J10 joinery redwood' or some other such quality. Only when the profiled and moulded joinery components have been machined from the timber and the joinery item assembled, can a decision then be made as to whether or not that finished item meets the BS EN 942 grading rules. So the best that may be said of any pack of 'joinery grade timber' is that it may have a reasonably good chance of producing an acceptably high yield of J10, or J20, and so on. Please remember that you cannot buy J20 (or whatever) wood directly from the importer.

I'm soon going to be tackling the tricky subject of Strength Grading, in the next chapter. But before I do that, I need to summarise what I've covered so far in this one.

5.29   Chapter summary

I started off by telling you that 'Grade' and 'Quality' are often used interchangeably: but that I prefer to use the term 'quality' to mean the timber's 'fitness for purpose'; whilst its 'grade' is the particular appearance of any individual piece, based on its selection according to prescribed criteria or published rules.

I then discussed the several variations on the theme of 'appearance' grading: covering first of all the Scandinavian and North American 'defect' systems, as used for grading Softwoods for general purposes sold in a range of dimensions. And I explained that Scandinavia and Europe no longer offer any 'clear' (i.e. fully defect-free) grades; whereas the USA and Canada can still offer some high quality, essentially defect-free grades, at the top end of their Quality Scale.

I also advised you that Scandinavia has a somewhat daft name for its best quality: which is a term called 'Unsorted', and which seems to be in defiance of any English language norms, since it actually means that it *has* been sorted or graded into the best available selection. Below that, they have grades called 'Fifths' and Sixths' which have evolved out of their system which has been in existence for well over 100 years. And I then told you that the Russians have something ever-so-slightly but just a little confusingly different: because they have 'Unsorted', and then 'Fourths' and 'Fifths' qualities.

The USA and Canada have some very different (and usually much larger diameter) Softwood tree species, for which they have independently evolved their own grading rules. And these 'Export R List' grades give us such things as 'Clears', Merchantable' and 'Commons' – in various subdivisions of better or lower appearance qualities.

I then explained that Hardwoods – both Temperate and Tropical – are graded using an entirely different 'appearance' method, called the 'cutting' system. This method relies upon the end-user being able to cut out the defects he doesn't want, whilst still leaving him with a minimum specified percentage of useable 'clear' timber, in a range of minimum useable sizes. The rules used for most Temperate Hardwoods are called the NHLA Rules and originated in the USA; whilst the rules used for the majority of Tropical Hardwoods are called the Malaysian Grading Rules (MGR), which originated in the Far East.

And then, don't forget that each Shipper will have his own interpretation of the particular rules that he is notionally working to: so you should really try to gain some experience yourself, by looking at the timber that you may be offered, under the various names or Quality descriptions; then you will understand much better what these terms all mean in reality.

Finally, I said that if you are specifying or manufacturing joinery items, and using BS EN 942 as your reference, then you must be aware that the 'J' Classes within that Standard apply to finished joinery only – and not to packs of timber that a manufacturer may buy from an importer or merchant.

6 Strength Grading and Strength Classes of Timber

Strength Grading is the primary example of what may be more accurately known as 'end-use' grading. In other words, it is another kind of selection process, but one that is very different in its intended outcome from that which I've just covered at length in the previous chapter. Strength Grading is, however, still based on the assessment of permissible defects within each piece of timber: but the fundamental difference here is that the piece of timber, at the end of the grading process, is intended – most definitely – to be fit for one very specific purpose.

It sounds straightforward, but that is a somewhat revolutionary concept in timber grading: the notion that a piece of wood can be sold as actually being fit for any particular job or end-use! What that really means is, after many decades of reluctance to commit to any specified uses for their products, the Timber Trade have (somewhat grudgingly at first) agreed to endorse a method of producing timber that can be used directly for building with (see Figure 6.1).

So, the particular end-use of any piece of Strength Graded timber – as that name so obviously implies – must be something that has to do directly with the *strength* of the timber. And all of the graded pieces are intended to be used *as they are*, without any further processing or reducing their dimensions significantly (other than perhaps cutting to length). And it is always expected that they will be used for load-bearing, structural members – such as beams, joists, posts, studs and roof timbers.

Older readers – perhaps even you, yourself? – may have come across the term 'stress grading': and you may possibly be wondering by now just where that name fits into the scheme of things. Well, dear reader, I can tell you that 'stress grading' and 'strength grading' are (or rather, were) in fact one and the same thing.

When I took my first timber grading exam, back in the mists of time (in the Summer of 1975, to be precise), the process was indeed always known as 'stress grading': for the perfectly sensible reason that the timber was graded in order to be able to resist 'stresses'. In other words, applied forces or loads – such as compression, tension and shear – that the timber member could be subjected to. But somewhere along the way – around the late 1980s or early 1990s, as

Wood in Construction: How to Avoid Costly Mistakes, First Edition. Jim Coulson.
© 2012 John Wiley & Sons, Ltd. Published 2012 by John Wiley & Sons, Ltd.

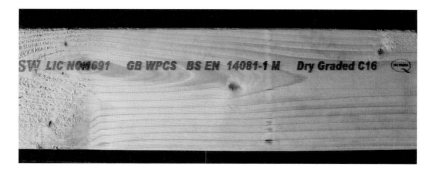

Figure 6.1   Typical strength grading mark on structural timber

I recall – the UK adopted the European way of describing things; and the terminology was changed. So it then became known as 'Strength Grading'. This was meant to show the users very clearly that the intended purpose of this particular grading method was to select timber that was structurally strong (i.e. graded specifically for its strength characteristics); so that it was capable of carrying 'engineered' or 'designed' loads.

And that is a very important distinction to make, in terms of the end-uses of these particular timber components. A piece of Strength Graded timber is of course meant to be strong: but that is *all* that it is required to be. It is not meant to *look nice* as well! So, the general impression of all those various visual 'defects' which only affect the 'nice' look of a piece – and which I was considering in some detail in the last chapter – should really not bother the buyer or seller of Strength Graded timber very much: as long as the wood is adequately strong for the job it was designed for.

I repeat: those defects which affect appearance should not bother the customer at all. In other words, a 'pretty' appearance shouldn't really be a consideration, when it comes to Strength Graded wood. So long as the timber is strong enough to do its job, then appearance really (really!) doesn't matter a bit. But unfortunately, here in the UK, it does seem to!

I work closely with sawmills in Scandinavia and Europe, helping to both train and Certificate their Strength Graders so that they can export timber to the UK which meets all the relevant rules. And I am always being asked to take the same old message back home to the UK with me: 'Why do your buyers ask for limited blue stain on C16, and no blue stain on C24?' In other words, these overseas mills cannot understand why our Timber Trade – and their customers – are refusing to accept something that is purely cosmetic and which the rules plainly allow. And the straight answer is ignorance – combined with a very old-fashioned and entrenched attitude to timber.

I'm always being told by UK builders and other wood users: 'Timber isn't the same these days: it doesn't grow like it used to.' Or a variation on the theme: 'You can't get the quality of timber nowadays that you could get when I were a lad,' etc, etc. And in a way, they're right: but in another way, they're very wrong.

Of course, the quality and availability of timber *has* changed over the past 50 years or so: but not always for the worse. Forest practices are much better now than they were before. Even in Scandinavia and Western Europe, where their forest management was always first rate, there has been greater attention given in the past couple of decades, to the environmental and social aspects of growing more and more trees for ever and ever (which of course is eminently possible). And in the Tropics – where the picture is much more patchy, and forest management practices are only slowly improving – great strides have been made more recently towards more local sustainability and globally-recognised Certification. But it is, however, true to say that we can no longer buy all of the large sizes and long, clear lengths that we once could. And that's perhaps no bad thing.

It was quite normal, in our grandfathers' or great-grandfathers' days, to be able to buy 300 mm by 300 mm Baltic Pine beams, almost knot-free. And as I have been explaining in the previous chapter, that sort of wood is simply no longer readily available to the Timber Trade. But there in a nutshell is the whole point of grading: to be able to select particular qualities of timber that will be suited to particular jobs, without being any better or any worse than they absolutely need to be. So in that way, we can much better conserve our forest resources, all the way to infinity – or as near to infinity as Mankind might ever get.

6.1   Appearance versus strength

Therefore, it ought to be perfectly acceptable for any wood user to expect, or to be offered, timber that is 'just right' for the job – *but no better than it needs to be.* And that's where the UK has, I'm afraid to say, missed the whole point of Strength Grading. So long as the wood is strong enough for the job, and its appearance is not important to the finished product (that is, when it is used only for the structural framework of a building), then the 'look' of the timber which is used in that building is – or it certainly should be – more or less irrelevant. Just think about that for a few moments.

The next time you are in a completed building (any building, that is, which doesn't have any special architectural timber features), have a close look at the structural timbers. Or rather I should say, have a look *for* the structural timbers. Can you see the floor joists? Or the roof timbers? Or any of the load-bearing studs within the timber walls? Can you? Of course you can't! Because they are hidden under the floorboards and the carpet; or above the bedroom ceilings; or behind the wallpaper and the plasterboard.

So why on earth should it matter that the joists or the studs are blue-stained, or that the roof truss timbers have a few pinworm holes in them? But if, on the other hand, you were to fall through the floor, because the joists broke under your weight due to the presence of excessively large knots; or if the roof were to cave in because the truss timbers had rot patches in them, then of course that would matter a great deal. But to complain about any of the structural members being 'ugly' or 'unsightly' when you and your clients will never see them once the building is complete, is surely missing the point!

And yet builders do precisely that, all the time. But in so doing, they are – and I'm sure, mostly unintentionally – wasting a huge amount of perfectly usable timber: or at the very least, relegating it to some lower-class and lower-value uses. What a daft state of affairs that is: no wonder the Europeans think that, when it comes to what they find acceptable in terms of Strength Grading, the UK Timber Trade is completely bonkers!

6.2   Visual strength grades

There are various Strength Grades available within Europe, all of which are selected – by eye – by highly-trained personnel, using specific rules to assess the various strength-reducing defects that occur in the lengths of timber they are grading. The two most common sets of rules used for visually grading Softwoods are BS 4978 from the UK, and the Scandinavian INSTA-142 Rules for Nordic Timbers. I will leave aside the latter, for the purposes of my present discussions in the context of the UK market: although they are nevertheless used quite a bit when trading Strength Graded timber between different European countries on the Continent.

But BS 4978 is the main Visual Grading Standard that most exporters will use when selling timber to the UK: and it is the UK that is by far Europe's largest net *importer* of timber and wood-based products.

6.3   GS and SS strength grades

BS 4978 specifies the grading rules for producing two distinct Strength Grades: General Structural and Special Structural – known universally as GS and SS, for short. GS is the lower grade, used for most 'bog standard' construction; whereas SS is the higher grade, intended to be used for longer spans or higher loads; or to enable smaller timber sections to be used. I said 'intended' deliberately just now: because another rather silly situation has arisen in respect of these Strength Grades, and the Strength Classes they will give rise to (see the explanation of their inter-connectivity a bit later on in this chapter).

For the same (misguided) reasons that I just went on about, with regard to the fundamental misunderstanding between Appearance grades and Strength Grades, it seems to have now become the norm for architects and designers (but *not* structural engineers, I'm pleased to say) to ask for the higher structural Grades or Classes, simply on the basis that *they will look better*. But of course, as you now know – because I've just told you in no uncertain terms – that whole notion of looking better because it is stronger is simply codswallop. SS grade is just one of the stages in producing structural timber which will be somewhat stronger than timber of the lower, GS grade. But *it still doesn't have to look any better*!

The grading rules contained within BS 4978 permit blue stain to be present *without any limit* in both GS and SS grade: <u>because it is a structural grade and not an appearance grade</u>. (Sorry to resort to underlining, but that very point seems to be completely lost on architects and designers, as it has for the past 30 years – but it is really important.) GS grade will, of course, contain more of the important, strength-reducing defects (large knots, steeper grain angle, splits, etc) than will SS grade: but that's the only difference between them. They could both look clean, bright and square-edged; or they could both look blue, wormy and waney-edged. And either appearance – pretty or ugly – would still be perfectly valid under the BS 4978 rules.

However, simply asking for GS or SS will not help a structural engineer to design a beam or a rafter, or whatever. The engineer needs to have some numbers (which are known as 'characteristic values' or something of that sort) to work with, and to put into his or her calculations and design formulae. So what the structural engineer needs to know about, and which some inspector or someone, needs to look for evidence of on the wood itself, is the timber's Strength Class.

6.4   Strength classes for softwoods

I mentioned C16 and C24 a short while ago; so I'd better explain what they are and where they fit into the scheme of things. When we grade timber, as I've just been explaining at some length, we select it on the basis of allowable defects which are permitted within the rules. But the *grade* of timber, as I've also been explaining, doesn't help, on its own, to decide how strong that piece of wood is. But then neither will only knowing the wood *species* help the designer, either. On a comparative basis, we can say that such-and-such a species is generally stronger than another; but we cannot say precisely *how* strong any given piece of timber is, just by reference to its wood species alone.

What an engineer needs to know is a *combination* of those two salient facts, so that he or she can get some meaningful figures, to put into a set of 'proof' calculations for a particular member or part of a structure. Remember, it's the *numbers* that are important to the engineer: and those numbers, so far as timber is concerned, are obtained via the Strength Classes. And these Strength Classes in turn, as I hope is now a bit clearer, are obtained from a combination of both the wood species and the timber's particular Strength Grade. Happily, there is another European Standard that neatly sorts all of this out for us: and it is EN 1912.

6.5   BS EN 1912

This Standard – in its British version – lists all of the European (and indeed, those of other countries outside the EU) grading rules, plus the Strength Grades which those rules contain. It then compares all of those grades with the different timber species that may be sold within the EU (even, as I have indicated,

Table 6.1   UK species/grade combinations and the resulting strength classes, adapted from EN 1912

Species	Grade	Strength class
Douglas Fir	GS	C14
Pine	GS	
Spruce	GS	
Larch	GS	C16
Douglas Fir*	SS	C18
Spruce	SS	
Pine	SS	C22
Douglas Fir*	SS	C24
Larch		

*Douglas Fir graded SS and in section sizes exceeding 20 000 mm^2 rates as C24. Sections sizes up to 20 000 mm^2 rate C18.

including those which originate from outside the EU, such as Douglas Fir or Western Hemlock from North America). The tables within BS EN 1912 tell us which species and grade combination will result in which Strength Class (see Table 6.1 for the species/grade combination for UK-grown softwoods).

The aim behind introducing Strength Classes was to make life much simpler for the timber specifier. Instead of trying to second-guess the market place as to which grades and species might be on sale at any one time, or which might be cheaper or more readily available, a specifier would simply have to ask for a Strength Class within his or her design and then leave it up to the Timber Trade to provide a species/grade combination which satisfied that Class. If only it were that simple!

6.6   SC3, SC4: C16 and C24

The UK construction industry has got itself stuck (as usual!) in a historical time-warp. Some years ago, the UK Building Regulations contained a set of timber Strength Classes, numbered from SC3 to SC9; and which covered both Softwoods and Hardwoods. They were not, it may be said, a universal success. But after some years, the builders (and yes, even 'White Van Man') became used to asking for SC3 or SC4 for their floor joists or whatever. This was assisted very much, it must be said, by a set of Span Tables that were given in the Approved Documents to the Building Regulations, at that time. And these Span Tables existed for various load-bearing members – floor joists, wall studs, and so on – which used either SC3 or SC4 as their base requirement, depending upon the span and load. And all seemed to be working, until Europe intervened.

Another European Standard – EN 338 – was introduced in the early 1990s (it has been revised and extended a couple of times since: the latest version being BS EN 338: 2009). This gave us a new set of Strength Classes for Softwoods, numbered from C14 to C50: and in this document the *nearest equivalents* to SC3 and SC4 happened to be C16 and C24 respectively; although there are many other Strength Classes to choose from, including at least three that fall in between our standby ones of C16 and C24.

But therein lies the problem: the nearest equivalents to SC3 and SC4 turned out to be C16 and C24; and because of this, the builders (and the Timber Trade too, it must be said) ignored all of the remaining possibilities and focused their attention on just those two Strength Classes, to the exclusion of all others. 'So what?' you might say. 'So wasteful,' say I. Especially where the UK's own home-produced timbers are concerned.

Using the species/grade combination criteria from BS EN 1912 that I mentioned a little while ago, and referring to Table 6.1, you'll find that British Spruce graded to GS, can only give us C14: whereas imported European whitewood in GS Grade gives us C16. Now, why should that be? They're both Spruce, after all.

I'll tell you why. Because the growing conditions, soil, climate, and so on in the UK, mean that the wood itself turns out to be inherently *weaker* than the same wood species grown in Europe or North America. And it's just the same with Scots Pine versus European redwood: the former giving C14 from UK sources and the latter giving C16 from European (and Scandinavian) sources, when both are graded to GS.

I won't waste your time – or my page space – by going through numerous examples: I'm sure you get the drift. But please have a look in BS EN 1912 and see how many other anomalies you can find, between British Grown and imported species, in both GS and SS; where the same *Grade* results in a different *Strength Class*, when the wood species (or at least, its place of origin) is different.

My point here is that, by concentrating exclusively on C16 and C24, both our Timber Trade and our Construction Industry have completely scuppered the notion of the flexibility of Strength Classes in relation to using timber, as was originally intended by EN 338 and EN 1912. The UK seems to be 'locked in' to asking for only C16 and C24, when it could be using many other species/grade combinations much more economically, without wasting those which give lower or different combinations. Now here's a puzzle for you. If the UK will insist on having C16 all the while, then how does the UK Sawmilling Industry manage to produce only C16 British Spruce, when according to BS EN 1912, it should give us C14 from GS grade and C18 from SS grade?

6.7 Machine grading

The answer is: by putting the timber through a grading machine! There are several types of Strength Grading machines on the market, each using a variety of techniques, ranging from 'proof bending', to X-rays, to ultrasound

Figure 6.2　A typical strength grading machine

(see Figure 6.2). But the precise method by which these machines do their clever stuff is not the point here. This is not – as I said in the Foreword – a highly technical text-book. ('Well, you could have fooled me,' I hear you muttering under your breath.)

The essential idea to grasp here is that these Strength Grading Machines work, because they use test data obtained from hundreds and hundreds of tests done on different species and section sizes of timber, in order to create a 'characteristic value' for the strength of that wood species. And they then use that data to derive the Characteristic Grade Strength Values for each timber type, to enable them to be fitted into specific Strength Classes. So any machine (they are all computer-controlled) can be set to the particular parameters needed to proof-test any timber member of a given species and in a specified section size, in order to see if it meets a selected Strength Class.

On the basis of each proof test, any piece which falls below the machine's in-built criteria for the selected Strength Class is regarded as a 'reject' and anything above that is a 'pass'. And it will be given a 'pass' even if its actual strength is way, way above the set Strength Class boundary. The overriding criterion is that no piece must fall *below* the minimum accepted level of strength, as denoted by the Strength Class to which it is being tested.

Thus, although GS British Spruce naturally gives C14 as its highest allowable allocation, there will be many pieces in a typical parcel which will be stronger than that absolute C14 minimum. Therefore, setting a grading machine to C16 will then simply 'cull' out all of the lower-strength pieces – i.e. those which are just below C16 - even if that culling (known, of course, as rejecting) includes many pieces that are actually C14 or better (but which are not quite up to being C16).

I hope you can now see why I believe that the UK's 'love affair' with C16 and C24 is so very wasteful. There is nothing 'wrong' or 'illegal' about designing a timber structure in C14, or C18, or C22 – all of which are obtainable from UK-grown species – but nobody does it, *because they don't know any better*. Perhaps you can help me to change this wasteful attitude, from now on?

6.8   Other strength grades: Europe and North America

I hinted earlier on that there are other sets of rules – such as the INSTA 142 Rules for Nordic species – which can give other grades: and all of those can be fed into BS EN 1912 to give usable species and grade combinations in the form of many different Strength Classes. But there are other Strength Grades, which we rarely see at present, mostly owing to economic factors, but which we used to get in the UK all the time, up until just a few years ago. These are the North American Strength Grades.

6.9   Select structural, No. 1 and No. 2 structural and stud grades

All of the above are visual grades, which are in everyday use in the USA and Canada, and which sometimes find their way in to the UK, if the market is favourable (due to price, or size availability, and so on). And they are perfectly legal and valid – indeed, they are also included within the Tables of BS EN 1912 and are allocated into Strength Classes, just as with the European grades. Naturally, these Strength Grades relate solely to North American wood species, such as Douglas Fir, Larch, or Western Hemlock. They also operate for Species Groups – for example, Spruce-Pine-Fir (more usually known as 'SPF' for short). This is a species combination which is sold as one parcel, where the individual species are not separated out.

Unfortunately for the North Americans, these species/grade combinations do not often work out favourably when being squeezed into the template of European Strength Classes. So you will find, for instance, that Select Structural grade Douglas Fir has to be slotted into C24, when it is in reality a bit stronger than that category would indicate: but there is no intermediate Class into which it can be put. And the situation is even worse with regard to No. 1 and No. 2 Structural. They are deemed (under this European system) to be so close in strength, that they are both lumped into C16: and yet in the USA or Canada, their strength differences are sufficient for them to be separated out and used individually.

For this reason, many North American producers decided, back in the days when the 'C' Classes were being introduced, that they would train their graders to use the UK grades of GS and SS, so that their timber species would be more fairly allocated into C16 and C24 on their own grade merits; thus making their production more cost-effective. It is a great pity that the present UK timber market situation – with its cheaper UK and European supplies and its high Dollar/Pound exchange rate – has meant that all the Canadians' efforts at 'converting' to our rules have been somewhat wasted, at the present time. Nobody has bought much North American Strength Graded timber for the past 15 or 20 years: but then, we in the UK are part of a much bigger world market, so things can always change – especially the Dollar/Pound exchange rate. So you may find that, by the time you buy and read this book, you are either buying or being offered Canadian SPF, graded to GS grade and allocated to Strength Class C16. (Then again, of course, you may not.)

6.10   TR26

'What's TR 26?' you may well say. And if you're not concerned with making or specifying trussed rafters (see Figure 6.3), then you can skip this bit.

In the past – as I have recently explained – there were 'old' Classes known as 'SC something-or-other'. And within these, there was a further, higher, Strength Class which got used – but only by the trussed rafter industry – and which was called SC5. (Which itself was satisfied by an obsolete machine grade, called M75: but let's not go there!) Now, SC5 was used for engineered roof truss designs on account of its higher strength, because this enabled truss

Figure 6.3   Newly manufactured trussed rafters

designers to produce complex, highly engineered designs which nevertheless utilised relatively small sections of timber. Then what happened when the 'new order' came in, was that SC5 was substituted by a higher Strength Class, called C27.

However, the fly in the ointment was that, when suppliers of the former SC5 used the new settings for C27 on their grading machines (SC5 could never be visually graded, by the way, since there was no visual grade high enough to reach it), they found that they were suddenly getting lower yields of Strength Graded timber from the same material that they'd used for SC5. Shock, horror!

But then – to cut a long story very short – after much head-scratching and scrutinising of the rules and regulations, it was discovered by a member of the UK Timber Grading Committee, that any EU member country was permitted to create its own Strength Class of timber, to be specified for a particular end-use. So, after a classic bit of 'British fudging', the machine settings for SC5 were simply re-issued as TR26 (the letters 'TR' of course standing for 'Trussed Rafters' and the number 26 showing that it is just a bit less strong than C27). So at a stroke, the UK trussed rafter industry was given back its former Strength Class and the related production yields. As I have just said, the letters 'TR' refer to the designated end-use of this 'special' Strength Class, as being intended only for the manufacture of Trussed Rafters: but I have begun to see TR26 used for floor joists and other structural members as well, apart from roofs. Whilst that is probably not unsafe; it does rather go against the permission under which TR26 was granted to the UK in the first place.

Of course, the trussed rafter industry could easily – and perfectly legitimately – design its trusses using C16, or C24, or C30 (or even C27 as originally intended!): but, as I'm so often complaining, we do not have the most progressive or technically-minded of workforces, insofar as the UK Timber Trade or the Construction Industry are concerned. So they have preferred to stick with what they already knew, rather than try out anything else.

6.11   Specifying the strength class or the wood species: some things to think about

I have repeatedly said that the thinking behind the introduction of Strength Classes was to make life somewhat simpler for the specifier and to leave it to the timber supplier to come up with a species/grade combination which met the specified Strength Class. And by and large, that's still what should – and often does – happen. But there are times when that may not be the absolutely correct thing to do.

If a specifier *only* asks for a Strength Class, then it is quite legitimate to supply anything which does that job, strength-wise. But sometimes, specifiers may actually *name* a particular wood species: and in such cases, the onus is on the potential supplier to ask whether a substitute species would be either acceptable – or indeed appropriate. Let me explain.

Say a builder orders some C16 floor joists, but states that he wants European redwood. The typical stocks, in a Timber Merchant's yard at the present time, will be either C16 British Spruce or C16 European whitewood (which consists primarily of another species of Spruce); both of which are available in the usual 'carcassing' quality. (Carcassing is the general term given to either rough-sawn, or hit-and-miss planed timber, used for building: as opposed to the higher appearance qualities normally used for joinery items.)

So the only European redwood that the Merchant is likely to have readily in stock, is probably going to be some Unsorted Scandinavian redwood, which he usually sells for joinery uses; and which is therefore going to be more expensive than the Spruce carcassing timber that he has lots and lots of. But still, the Merchant needs to ask himself – or preferably the client: 'Why do you want redwood for this particular job?'

Perhaps the builder simply likes the overall appearance of redwood: or maybe he's an old-fashioned type, who's used redwood for 30 years and doesn't want to change. Or perhaps the floor joists are to be used as replacements in a house which has suffered from Dry Rot: in which case, they will need to be treated with high levels of preservative to Use Class 4. (And you will know from Chapter 4 that Spruce is notorious for being difficult to treat to Use Class 4: so it is not the ideal candidate for use where there has been a history of severe decay. The best timber in this case is European redwood – which of course, is a Pine – because it takes preservatives very well. You remembered that, didn't you?)

So, the builder (or whoever) may have chosen his named wood species for any one of several reasons: therefore supplying just 'any old C16' may be simply not good enough. You need to ask, before you impose your own change on the specification. I hope that's clear?

6.12   Hardwood strength grades

Now I'd better turn to the structural use of Hardwoods.

What I didn't tell you earlier on in this chapter, was that the letter 'C' in the 'C Classes' doesn't actually stand for the word 'Class'. Whilst the term 'SC' in the UK did indeed mean 'Strength Class', in this case the 'C' actually stands for 'Conifer': and that's because all of the European 'C' Classes relate only to Strength Graded Softwoods. Therefore, it is just not possible – as I have sometimes seen in orders or specifications – to get hold of (say) 'C16 Oak': there is simply no such thing (because Oak is a Hardwood, of course).

So what's the situation regarding the grading of structural Hardwoods? As you may by now have expected, there is indeed a range of separate Strength Grades for Hardwoods, all of which are given in another British Standard: BS 5756. And although it is only a single British Standard, it holds within it two separate sets of grading rules: one for Tropical Hardwoods and one for Temperate Hardwoods.

There is only *one* grade for Tropical Hardwoods, called 'HS' grade; but there are *four* grades for Temperate Hardwoods: known variously as 'THA', 'THB', 'TH1' and 'TH2'. I'm going to deal with Tropical Hardwoods first, since the situation there is somewhat more straightforward.

6.13   Tropical hardwoods

Of course, not all of the Tropical Hardwoods are used structurally. Only those with particularly good strength properties (naturally!) and, by and large, those which do not have a particularly notable decorative appearance, are included in the structural category. But there are exceptions: Iroko and Teak are both Tropical Hardwood species, each of which has a good reputation as an attractive and versatile joinery timber, with an appealing character (known as figure, if you recall that from the earlier chapters of this book). Both of these species are also recognised as having good structural potential as well. And in many instances where specific Hardwoods are concerned, it is very often *both* their appearance and strength characteristics which will have influenced the choice of that particular timber for structural use: where perhaps a featured beam or roof truss is intended to be exposed in a building, and the architect or designer wants the particular figure, texture or colour of a certain type of Hardwood to be on show.

The reason why Tropical Hardwoods are only allocated a single Strength Grade, is that they do not vary overmuch in their characteristics: they are largely free from knots, and their grain is often very consistent, albeit somewhat different in nature from the Temperate timbers. The main reason for the grading process is essentially to ensure that 'rogue' or 'weak' pieces do not slip through the net.

Even so, most of the structural Tropical Hardwoods have quite different characteristic strengths from one another: so once again, it is necessary for designers or engineers to put together the particular wood species with the single HS Grade, in order to find out what the relevant Strength Class is, on which to base their calculations.

6.14   Temperate hardwoods

Let me say clearly and unequivocally, here and now, that there is no such thing as 'Architectural Grade' when it comes to structural timbers! Even though some such fancy-sounding name has been bandied about – especially in respect of Oak – for many years: it is complete codswallop. In fact, it is worse than that: it is meaningless and potentially dangerous codswallop! *All* structural timbers, by law (i.e. the Building Regulations) *must be* Strength Graded if they are to be used for load-bearing applications in building: and that applies just as much to the older, 'traditional' Hardwoods as it does to the newer and more 'common' Softwoods.

The situation with Temperate Hardwoods is more or less the opposite to that which exists with their Tropical counterparts. Only a handful of Temperate Hardwoods – Oak and Chestnut from the UK, plus White Oak and a couple of other species from the USA – have been fully tested for strength, and have therefore had any reliable design figures (stress values) published for them.

As I mentioned briefly in the introduction to this section, there are *four* Strength Grades for Temperate Hardwoods; and this is because broadleaf trees can often show much greater variation in the defects they contain: such as knots (from very large branches or from groups of very small 'adventitious' shoots); straightness of grain, or other strength-reducing characteristics. Once again, when checking what design figures to use in an engineer's calculations, it is a matter of combining the particular Temperate Hardwood species with a relevant Strength Grade – as selected from the four possibilities available, to arrive at the requisite Strength Class.

I'll tell you what those Hardwood Strength Classes are shortly; but first, I want to tell you about a peculiar phenomenon in respect of Temperate Hardwoods, which has resulted in the need to create those four distinct Strength Grades.

6.15   The 'Size effect'

When the Building Research Establishment (BRE) was testing the hundreds of samples required to build up a bank of strength data, from which to extrapolate both the characteristic strength, and to set the grade boundaries, for a particular wood species, they found that a strange thing seemed to be happening. What the BRE boffins discovered was that, *all else being equal*, <u>larger</u> section sizes were showing up as being stronger than <u>smaller</u> sections. This was nothing to do with the fact that a larger piece of timber can (of course) carry a greater load than a smaller piece: no – it turned out to be a fundamental property of such large-section timbers, that they are just *stronger*, per se. Now, how odd is that?

On checking the results from all of their tests, it appeared to be a consistent phenomenon, that any piece of timber with a cross-sectional area greater than 20 000 mm² (that is, about 8" by 4" or equivalent) *and also* with a thickness of not less than 100 mm (4") was actually *stronger* than a piece of the same, identical grade, but which was smaller – in terms of either its thickness or its total cross-sectional area. Weird! That means, after so much speculation on the subject, that size *does* matter, after all!

So – joking apart - the 'size effect', as it is correctly known, is the strange but true reason behind the need to separate out four different Temperate Hardwood grades. There is a 'high grade' and a 'low grade' (which are more or less equivalent, in principle, to the idea of GS and SS in Softwoods, if you will: although their grading rules are different, of course). But there are also the 'large' and 'small' section sizes of timber to be considered, too. Thus, by combining the various permutations of 'high', 'low', 'large' and 'small', the chaps at BRE ended up

Table 6.2 Relationship between THA, THB, TH1 and TH2

	Large section	Small section
High Grade	THA	TH1
Low Grade	THB	TH2

*The large section refers to any timber that has a cross section of 20 000 mm²
or more, and no dimension less than 100 mm.
**The small section refers to any timber with a cross section of less than
20 000 mm², or thickness less than 100 mm.

**Table 6.3 UK Hardwood species and grade combinations for different
strength classes**

Strength class	Species	Grade
D30	Oak	TH1
	Oak	THB
D40	Oak	THA

Oak in TH2 does not meet any Strength Class.
Sweet Chestnut may be Graded to THA, THB, TH1 and TH2 but these Grades
do not meet any Strength Class.
Designers must use the individual Stress Values given in BS 5268: Part 2 or
the Basic Values from Eurocode 5.

with needing four possible Strength Grades for Temperate Hardwoods, to cover
all eventualities. Just have a look at Table 6.2 to see how they relate to one another.

6.16 Hardwood strength classes

I told you earlier on that the Softwood Strength Classes are designated with a
'C' for 'Conifer'; so you might expect something similar to be the case with
Hardwoods: and so it is – well, nearly!

The Hardwood Strength Classes are known as 'D' Classes: and they range
from D30 up to D70, the number indicating that they are considerably stronger
than most of the 'C' Classes. And the letter 'D' stands for 'Deciduous': which is
not strictly accurate, since there are many broadleaved trees that are *not* deciduous (practically all of those in the Tropics, for example); and then there is Larch,
which is a Softwood, of course, that *is* deciduous. Never mind: 'D' it shall be.

The main thing that I really need to emphasise here, is that *all* Hardwoods –
of both the Temperate and Tropical kinds – *must* be allocated into a Strength
Class by means of their individual wood species and Strength Grade
combination (see Table 6.3). That's because there is currently no machine on

the market, which is capable of proof-testing any of the Hardwoods and thus putting them directly into a Strength Class, in the way that can be done with the Softwoods.

6.17   The marking of strength graded timber

Even when the timber – either Softwood or Hardwood – has been properly graded and its relevant Strength Class determined, that is not quite the end of the process. There is a relatively recent European Standard – called EN 14081 – which governs all of the timber Strength Grading operations across the EU. Requirements for marking strength graded structural timber existed long before the publication of EN 14081: but these requirements were pretty well country-specific; and some EU countries (like the UK) were better at doing it than many others.

Nowadays, EN 14081 requires *all* producers of Strength Graded timber across the whole of the EU to follow the same procedures. Even timber graded and sourced from *outside* the EU must follow all of the EN 14081 rules, if it is to be sold in any of the EU's markets.

In essence, the rules in EN 14081-1 (that stands for Part 1 in current Euro-speak, by the way) advise that *every piece* of graded timber intended for structural use, should have certain information stamped indelibly, at least once on its surface. The exact format of that information can vary, since EN 14081-1 permits a large number of Options: but there must at least be something, in amongst the myriad symbols and initials on the stamp mark, which tells the informed purchaser or inspector what's going on. This can be either done directly – by giving the producer's name, the timber grade, and so on – or it may be done indirectly, via some code number or letter combination.

So each piece of timber will then reveal who produced it, who graded it, which rules were used; and what its wood species and its Strength Grade are. But the most important thing of all is that the stamp mark must *always* say what the timber' Strength Class is. In fact, even allowing for the potential for total obfuscation which could be caused, by using all of the various codes and letter or number combinations which are allowed, it is still (thankfully) the case that the Strength Class of each and every piece *must* be clearly marked upon its surface, at least once, whenever there is a Strength Grading stamp on the timber. Have another look at the illustration in Figure 6.1 and I will explain it a bit (see Figure 6.4).

In certain circumstances – and this is usually only permitted for aesthetic reasons, where the use of an indelible stamp mark would obviously mar the surface appearance of a timber member intended to be featured – the grade stamp mark may be left off the timber. But where this dispensation is allowed, then each 'batch' or 'parcel' of graded timber must have a Certificate of Grading accompanying it to its final destination, so that its 'grading provenance' can be maintained. The Grading Certificate must include on it everything that the

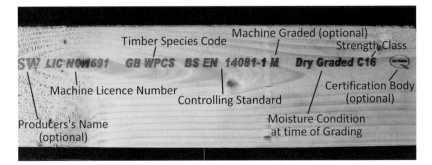

Figure 6.4   Typical strength grading mark on structural timber

stamp would have shown: plus additional information concerning the length and size specification of the timber parcel, plus the number of pieces within that parcel: so that in the event of a query as to the accuracy of the grading, all of the timber can be traced back to the source mill or producer. It's a simple but effective way of checking that the grading has been done and to then see whether it was done properly.

Now I need to tell you again what I've just told you, in my summary.

6.18   Chapter summary

I began by telling you that Strength Grading refers to a very specific 'end use' of timber. It is the notion of using a piece of wood for something very specific, rather than it simply being sold as not necessarily being fit for any particular purpose. And I made the point that *strength* is the over-riding criterion, with the timber's appearance generally being very secondary; if not completely irrelevant most of the time (that's particularly so when it comes to carcassing Softwoods).

I explained what is meant by a Strength Grade, as opposed to a Strength Class; and I said that structural timber may be graded either visually or by a machine – at least insofar as Softwoods are concerned. Remember, I told you that Hardwoods must always be visually Strength Graded, since they are inherently too strong for any machine to test properly.

I told you about the 'C' Classes which are used for Softwoods and the 'D' Classes which are used for Hardwoods; and the fact that these Strength Classes are different from Strength Grades – the Classes give the figures for engineers to use in their calculations. I also explained how Temperate and Tropical Hardwoods are different; both in respect of the Strength Grades given for them and in the way in which those Grades are combined with the wood species to achieve a particular Strength Class.

I explained that the most common Strength Classes used in the UK are C16 and C24: but I also told you that there were many others which can legitimately be used, even though they are not. And I also explained what TR26 is and how it came about.

Finally, I said that *every* piece of Strength Graded timber must either be marked or it must have a Certificate of Grading for its production batch. And the mark or Certificate *must* include information as to the timber's Strength Class.

So now I've got you to the point where you can better understand which timber to select, for which particular job. Now I need to look at a couple of things you might do with it: 'treat' it or 'paint' it.

7 Wood Preservatives and Wood Finishes

The first of many misconceptions about timber preservation – and one that needs to be put right without delay – is the almost universal myth, that by putting wood into some sort of tank and then impregnating it with some sort of chemical, it will be somehow 'treated' to the point where it is completely full of preservative throughout its entire length and breadth: and so then it will be immune to attack forever. Nothing could be further from the truth!

I hope that by now, you will know enough about the structure of wood to realise that it cannot – and indeed does not – act like a sponge and simply absorb everything that anyone might try and soak into it (see Figure 7.1). I've already covered the concept of 'permeability' in an earlier chapter, and I've explained that many wood species – including Spruce, as the most well-known common example – are extremely difficult to impregnate, on account of their particular cell structure. By the same token, the heartwood of pretty well all wood species, being 'closed down' and no longer taking part in the tree's growth, is much more resistant to impregnation with preservatives than is the sapwood zone of the same timber (Spruce and other timbers that are rated as 'resistant' excepted: since *all* of the cross-section of any such timber is hugely difficult to get anything into).

7.1 Treat the timber last!

The practical upshot of this simple but much-misunderstood fact about the difference in genuine permeability between different wood species, and also between sapwood and heartwood, is that in the real world, *not all of the piece of treated timber will contain preservative*. And the net result of that particular state of affairs is that, effectively, the timber is only being protected by means of a '*cordon sanitaire*' – or a protective layer – which sits as a sort of 'skin' around all of the at-risk parts of the component. It must therefore stand to reason that any breach of this protective layer, or barrier, will put the timber at serious risk of decay or insect attack. And it therefore follows that the worst thing anyone can do to a piece of treated timber is to deliberately make a cut into its cordon sanitaire, and by so doing, to seriously risk the ingress of something nasty into the unprotected part of the wood.

Wood in Construction: How to Avoid Costly Mistakes, First Edition. Jim Coulson.
© 2012 John Wiley & Sons, Ltd. Published 2012 by John Wiley & Sons, Ltd.

**Figure 7.1   Treated zones in spruce (upper) and pine (lower). Note the low
penetration into the pine heartwood**

Because of this important – but hardly ever fully appreciated – reason it is
always necessary to do any of the cutting, drilling, notching or other trimming
of a treated timber component *before* any treatment is done to it. And yet,
I have come across countless examples where treated timbers have been cut into
shorter lengths, or have had great big holes bored into them; and they have then
been put in the ground, or exposed to wetting in some other way. Only recently
(I'm talking here of October 2010) I was inspecting some Softwood timber
gate-posts that had rotted prematurely: and when I asked about their history,
I was informed that they were square sections which had been 'pointed' at their
bottom ends, just before being driven into the ground; and – of course, this is
the cruncher – that was *after* they had been impregnated with preservative.
Therefore, most of the preservative at the bottom ends of these posts had been
(rather unhelpfully) removed, as part of the wedge-shaped slices that were
taken off in making those nice, sharp points! So in the end, the cost and effort
of the treatment was all a bit 'pointless' you might say? (Sorry about that.)

7.2   Wood preservative types

I have already told you about the whole concept of 'Use Classes' in Chapter 4:
but at the time, I left out the exact details of the sorts of wood preservatives that
are available to help deal with some of the more hazardous uses of timber.

I simply made the point that any susceptible wood species would need to be treated with a suitable sort of wood preservative, if it was to be used in any hazardous situation. (Anywhere, that is, except for Use Class 1.)

But it is a fact that more or less every single use of wood, from Use Class 2 downwards, can result in the need for some sort of preservative treatment being done to the timber, to a greater or lesser extent. The only exceptions are those where the economics of the job may dictate that treatment is unnecessary (such as in a temporary building, perhaps), or where a particular wood species which has an appropriate degree of Natural Durability has actually been specified and used. Remembering, of course, that its sapwood needs to be fully removed – otherwise some additional preservative treatment will still be necessary.

So, how does the preservative get into the wood? What types of treatment are there? And are they all the same as each other? And can they all be used to satisfy every Use Class? I will now elaborate.

7.3   'Old' and 'new' types of treatments

The situation today is, in some ways, a good deal more complicated than it was a few years ago: although in other ways, it is perhaps more straightforward. There is no doubt in my mind that the concept of Use Classes has greatly simplified matters, so far as the specifier is concerned. But the greater variety of many 'new' and (chemically speaking) quite varied processes and treatments which have come onto the market in recent years, has led to a bit of confusion as to exactly which treatment or process can be used where. So I propose to tell you now about the two basic methods of application that are related to the various wood preservatives that exist.

7.4   The basic methods of timber treatment

Of course, it is possible to apply wood preservatives directly by brush or spray, just onto the surfaces of timber components. But the uptake and penetration of more or less all preservatives by such methods will be minimal, as might be expected (though not always, to judge by some peoples' expectations!). But for all practical and meaningful purposes, all wood preservatives, of whatever type, are impregnated into sections of timber by placing the timbers into an autoclave (usually called the 'treatment tank') and then encouraging the chemicals to penetrate into the timber by means of variously-applied levels of vacuum or pressure – or both.

There is one exception to this general rule about pressure treatment: and that relates to any of the Boron-based processes. But these have to be applied to unseasoned wood; and even then they are not well 'fixed' into the timber and so tend to leach out over time, when used in permanently wet situations.

However, they can be a good 'insurance' when used in wood that is normally dry, but which may get wet in the event of a one-off building or maintenance problem. I do not propose to spend any time describing them further here.

7.5   Low pressure treatment

This is commonly referred to as the 'double vacuum' method. It is the least penetrating of the treatments (the permeability of different wood species notwithstanding), since it relies on very little, or no pressure to push the chemicals into the wood surface – hence its title of 'low pressure treatment'!

There are two separate stages of vacuum applied during this type of treatment: one after the tank has been closed and sealed, and the other just before it is opened up again. The first vacuum is done in order to suck air out of the wood cells of the timber to be treated: so that there is space in the cell cavities and some in the cell walls to allow the preservative to enter the timber (air is easily compressed and so just putting fluid into the timber under pressure, without firstly drawing a vacuum, would simply squash the air inside the wood cells; which would then expand once more, as the tank was opened, and so push most of the chemicals back out again. A bit of a waste of time!).

After the treatment chemicals have been impregnated into the timber – by flooding the tank and then leaving the wood to soak (or 'dwell') for a while – a second vacuum is drawn. (Sometimes, such as with 'difficult' timbers like Spruce, a slight pressure may be added at the 'dwell' stage to encourage some better degree of penetration.) The purpose of the second vacuum is to ensure that any excess fluid is then drawn off; so that the timber surface is more or less touch dry, when it comes out of the tank. But, because little or no pressure is used, penetration of preservatives – even into the easy-to-treat species like Pine – is not very deep. For this reason, these low-pressure treatments are generally only suitable for Use Classes 2 or 3: dependent (as always!) on the wood species.

The chemicals used in these treatments are generally colourless; so it is quite common for treaters to incorporate a dye into the formulation, so that customers who require some 'proof' can then be reassured that the wood has indeed been 'treated'. This dye does not – of course – add anything to the longevity of the treated wood: but it perhaps gives a little reassurance to those who desire it.

The other thing about the low pressure treatments is that they use either solvents or – more frequently nowadays – low-water-content emulsions as the carrying medium for the preservative chemicals (this is so that they don't increase the moisture content of the treated wood). They are therefore very suitable for joinery products, where dimensional stability is most important.

7.6   High pressure treatment

This process is often referred to as the 'vacuum-pressure' method: since it has a high pressure element to the treatment cycle, which comes at the point in between the two vacuum stages. The reason for this extra period of high pressure is to try to force as much treatment fluid as possible into the wood cells during the treatment process, so that very high loadings of active ingredients and a good depth of penetration of the desired chemicals can be achieved. But – as I'm always cautioning you – the depth of penetration and the desired loading of active chemicals can only be achieved in wood species that are easy to treat.

With species that are classed as 'resistant', the penetration will be limited to a very narrow 'cordon sanitaire' around the periphery of the timber component, that I mentioned earlier. And in many cases, the loading of active ingredients may well be insufficient to cope with the most hazardous Use Classes – that is, UC4 and UC5.

The high pressure process uses chemicals that are first of all dissolved in water; and so this method is almost certain to raise the moisture content of treated components, at least for a time. For this reason, it is not recommended for joinery components or for any other precision-machined timber items. All of the high pressure treatments are based on some formulation that contains copper; so timbers treated with them naturally end up with a greeny-blue tinge to them. This is not a dye (as it would be with the low pressure treatments); it is a natural residual colour, caused by the particular chemical treatment.

7.7   Preservative chemicals

Before describing the newer chemicals that have come onto the market in the past decade or so, in response to environmental concerns, Id like to begin this part by saying a just few words in praise of the 'older' generation of Copper-based preservatives. These – despite a number of reports to the contrary – have not been totally outlawed, nor have they been completely withdrawn from use in Europe and the UK. (But you might have a problem still finding them!)

7.8   CCA preservatives

Salts of Copper, Chromium and Arsenic, dissolved in water – and better known to everyone in the wood world by their initials, 'CCA' – have been formulated into wood preservatives for over 70 years: and they have proven their value over that time as very effective protectors of wood from decay and, where necessary, from attack by various wood-eating insects.

Unfortunately for CCA and its many fans, it has been decreed by the environmental lobby in Europe, since the early part of this new Millennium, that the inclusion of that nasty-sounding and highly poisonous element Arsenic (and to a lesser extent, the undesirable Heavy Metal, Chromium) which are used within this highly tried-and-tested formulation, are 'unsuitable' for use in wood preservatives: at least as far as domestic uses of timber are concerned.

This ban was decided upon, despite there being a body of conflicting and inconclusive evidence about the likelihood of any leaching-out of these 'nasty' chemical constituents into soils or other areas either surrounding or near to any CCA-treated wood products. And this ban was enacted, despite the reams and reams of scientific papers and research studies proving how effective these chemical were. So the present situation is that CCA-treated wood has been banned from use in any domestic or similar application – such as children's playgrounds, animal fencing, and so on. But it may still be used perfectly legally in certain non-domestic and countryside uses; such as footbridges in National Parks, or some types of fencing in particular situations.

However, the extent of the wood-treatment industry's more or less willing 'conversion' to the newer chemical formulations has meant that, effectively, CCA as a treatment is just not available in the UK any more. CCA-treated timber may legally be imported here; but you will not find a CCA treater operating in this country. (And if you do happen to find one, I will be very happy to offer you my personal apologies, plus a heartfelt thank-you for finding it!)

So, the situation in the UK is that all the major preservative treatment companies have effectively given up on CCA treatment – because of the onerous restrictions on its use – and they will now only offer treatment with the newer 'environmentally friendly' formulations. So (like it or not) specifiers have been pushed towards the newer generation of wood preservatives, whose effectiveness is as yet relatively unproven (except by very short-timescale experimental laboratory trials); and certainly not from any actual long-term performance data from real life.

If you want some sort of assurance about the treated timber that you are specifying or buying, then you should do one (or both) of two things: you can ask for some sort of guarantee on the treated timber product – if the treater/supplier will give you such a thing – and/or you should specify treatment to a specific Use Class; as I have described *ad nauseam* in Chapter 4.

7.9 The 'environmentally-friendly' preservatives

There are a number of so-called 'environmentally-friendly' wood preservatives currently on the market. Many of them are based upon a group of synthetic, complex organic chemicals called 'azoles'. The leading chemicals within this group are Propiconazole and Tebuconazole: which form the major 'active ingredients' of a number of proprietary wood preservatives. Even the very well-recognised and trusted Brand-name wood preservatives – that is to say, the newer, more modern versions of the old CCA – mostly contain some sort of azoles as part of their chemical formulations, along with that most reliable ingredient: good old Copper.

Another family of wood preservatives – which still use Copper, I should note approvingly – is known by the acronym 'ACQ'. This stands for either (depending upon where you look it up!) 'Alkalised Copper Quaternary' compounds, or 'Ammoniacal Copper Quaternaries' (often abbreviated simply to 'Quats'). In any event, these are complex chemicals, whose formulations are based still on Copper; plus a selection of quaternary ammonium compounds. They were originally impregnated into timber using ammonia, although newer versions now use water as the carrier; but once again, ACQ's are not widely proven in service in the UK over long periods of use.

7.10   'Tanalised' timber

It's not my place to promote any one treatment or company: but I ought to at least point out that the name 'Tanalith' and its associated verb, 'to Tanalise' are brand names: currently owned by Arch Timber Preservation (who bought out the Hickson's Timber Preservation Company in the 1990s), and who are American and not Yorkshire-based, as Hickson's were.

But this brand name is so well known that many people use it as a generic term, without perhaps realising that it is indeed a brand name – in rather the same way that I will tend to use the term 'hoover' as a noun or a verb, when I should really be talking about a 'vacuum cleaner'.

The present 'environmentally friendly' version of the old CCA-based Tanalith preservative is now called 'Tanalith E' (for 'Environmental') and this uses a formulation based on Copper and azoles; but once again, it has not had the 60+ years of 'real life' in-service proof that its predecessor had.

7.11   Organic compounds

I should also tell you that other types of wood preservatives exist, which are in no way based on Copper. Instead, they are based on many of the new, complex, organic chemicals (such as the azoles which I talked about earlier); but they are often also combined with other chemicals – such as 'permethrins' – which are insecticides that will not harm bats. (You knew that it's illegal to kill bats, didn't you?) Unfortunately, the older types of insecticides, that were really highly effective against woodworm in old roofs, were reckoned to be responsible for killing off bat colonies: so out they had to go!

All of these newer organic-type chemicals are impregnated into timber using the low pressure method: and as a consequence, they are only effective for any treatments up to and including Use Class 3 (see Figure 7.2).

Figure 7.2   A typical high pressure treatment tank (photo courtesy of Osmose)

7.12 'Treated' timber

You should by now know enough to be very wary of the bald phrase 'treated timber', used simply on its own. That's because you also know that different timbers can have differing levels of uptake of chemicals; and you also know that the use of – or the lack of – a high pressure level during the particular treatment cycle must affect the uptake of whichever preservative is being used. So please, please refer back to Chapter 4 on Specification; and then make sure that you – or *someone* at least – quotes the appropriate Use Class wherever it is needed. Then make sure to have the timber of the correct species treated to that designated Use Class specification. Otherwise, the 'treated' timber simply won't last as long as you hope it will.

Now it's time to tackle the related – and equally misunderstood – topic of Finishes: especially where outdoor uses of timber are concerned.

7.13 'Wood finishes'

Even that very word 'finish' is capable of several degrees of ambiguity. Woodworkers and other craftsmen will often talk about the 'finish' on a piece of timber, when in many cases, they really only mean its level of surface preparation, or its surface smoothness. They are probably not referring to some type of oil or some other sort of clear or pigmented coating that may have been slapped onto the wood at some point in its life.

I could go on about paints and varnishes used on wood in indoor applications: but these uses are relatively straightforward and don't generally cause the timber too much trouble. So in this section, I intend to talk in detail only about the exterior finishes, that are (of course!) applied to wood out-of-doors: since that is the area where I so often come across the greatest number of problems, which are caused by ignorance and misunderstanding. And this is an area which would benefit, I think, from some clarity of understanding.

7.14 Wood in exterior uses

Wood used out of doors is subjected to a great many different things that wood used indoors does not have to suffer. These are things such as variations in temperature, and/or atmospheric humidity; and more especially, the vagaries of the British weather.

Of course, it has always been possible to use wood in an exterior situation (Use Class 3) without having to put any surface coating on it whatsoever. But in such cases, the wood will always go silvery-grey in a few years, as it 'weathers' down. This so-called 'weathering' is mostly due to a combination of bleaching-out of the natural wood colour by the UV light from sunshine, plus a build-up of trapped dirt beneath the surface fibres, which have become raised and also slightly deteriorated, by the actions of the sun and rain (see Figure 7.3).

Figure 7.3 Weathering of exterior wood without any protective finish

However, a wood species with good natural durability – such as Western Red Cedar – will not usually suffer from any decay or other form of degrade by weathering: it will simply fade away elegantly.

Although architects are sometimes known to favour this 'natural' wood look, I personally think that if timber is left without any surface protection, it will eventually start to look 'cheap and nasty' as well as a little unloved. And it also strikes me as faintly absurd that someone would spend good money on a timber like Western Red Cedar or Oak (as a cladding, for example) only to have the wonderful natural colour of that timber eventually lost to a drab greyness: a scenario in which any and every timber used out of doors without a protective finish, eventually ends up looking exactly the same as every other timber. Dull and grey.

7.15   Exterior finishes

So, in order to keep wood looking good out of doors and to help it resist the ravages of the elements for a reasonable number of years, it is necessary to conserve the natural colour of the wood with some kind of protective coating, or exterior finish. But please, please do not – whatever you do – use a *varnish*!

7.16   Varnish – and paint

I am at a loss to know why DIY stores still stock 'Yacht Varnish'. I've never heard of a yacht owner who pops down to his or her local DIY shop to buy a tin or two of varnish, when their zillion-pound yacht needs a bit of a touch-up! In my opinion, if such a product is not actually being used on a yacht, then Yacht Varnish has absolutely no place in the world of outdoor wood.

'If you can use it for yachts, then why can't you use it for cladding, or garden furniture?' I hear you say. But therein lies the answer. Yachts do not habitually spend several years out of doors, with no maintenance. They are looked after extremely well: they are taken out of the water at least once a year and scraped. They have crews whose job (when not actually sailing) will be to undertake 'routine maintenance', which includes scraping off and replenishing the 'tired' finishes on the woodwork of the decks and the hull. So the 'old' varnish never stays on the wood for years at a time; and it therefore has no time to break down and cause the problems of unsightly stain and mould, as it does when left un-maintained in your garden. That is why Yacht Varnish works: because it's on a yacht and not on a garden bench!

Normal paints don't really fare any better in outdoor situations, either. That's because, essentially, a paint is more or less a sort of 'pigmented varnish': and both paints and varnishes are what we Timber Technologists refer to as 'film-forming' finishes, which are really only suited to use *indoors* and not outside. All of the normal paints and varnishes in use these days have a formulation that is

Figure 7.4   Failure of film-forming exterior paint by peeling and flaking

based on some form of polymer (that is, a type of plastic resin); which as it dries, hardens or 'cures' to form a film or a skin on top of the decorated substrate. In the case where wood is that substrate, this plastic skin may keep its integrity for a year or so: but sooner or later, the plastic film breaks down, due to the effects of UV light and other weather-related stresses, resulting in the development of minute fissures (or microscopic cracks) in the surface of the paint or varnish. These fissures are, initially at least, far too small to see with the naked eye: but they are nonetheless present, and they will let in water.

You know by now of course that wood has a 'grain' structure: and that wood cells are very much like small tubes. So it should now be easy to understand that, when water enters the wood's surface via a crack in the paint or varnish coating, that water does not remain on the surface at the point where it entered: far from it. The water will run along the wood grain, by means of that wonderful physical process called 'capillary attraction' (the same thing that makes kitchen towels or blotting paper work so well), and then of course it ends up in a different place in the timber, *away from the point where it entered*. And by this means, the water often ends up trapped behind a more solid part of the plastic 'skin' that comprises the surface coating; and it then tries to get out, by means of evaporation. And in so doing, it then pushes off the paint or varnish in this new spot and creates further cracks: thus extending and prolonging the agony (see Figure 7.4). And of course, you also know that wood expands across the grain when it gets wet: so the expansion caused by the newly-raised mc in the wood's surface further disrupts the film and then either enlarges the existing cracks in it, or creates new ones: and so the water keeps on getting in, all of the time. But that's not all.

Long before the newly-wetted wood can lose enough of that trapped moisture back into the atmosphere – by disrupting the finish – it will raise the moisture content in that locality by a considerable amount: usually well in excess of the 20% safety level. And of course, that first of all results in the inevitable consequences of staining; and then eventually, you will get localised decay of the wood in that area. So much for the 'weatherproof' properties of Yacht Varnish, eh?

In fact, from a pure Wood Science point of view, I would much prefer there to be a flaky, peeling finish – or indeed, no finish at all – to a conventional paint or varnish: because at least in the places where the paint has come off altogether, the wood can finally manage to breathe and dry out!

7.17 'Microporous' exterior stains and paints

There has been a whole generation of so-called 'exterior' finishes around for about 40 years; but they have only really achieved any level of popularity in the UK in the past decade or so. They can be divided into two basic sorts: Wood Stains and Exterior Paints. Essentially, though, they work via the same basic physical mechanism: that is, they are 'microporous' – or if you like, they 'breathe' as the common parlance has it.

The original family of exterior wood stains was developed in the 1960s at the US Forest Products Research Laboratories in Madison, Wisconsin, using a concoction of boiled linseed oil, resin binders and pigments. Pretty well all of the modern exterior wood stains can trace their development back to this so-called 'Madison Formula', which has been modified and improved over the ensuing decades; but which is essentially the same thing, at the heart of all successful exterior finishes. (In Scandinavia, a similar but unrelated family of finishes has been around since the end of the 19th century: based on iron oxide – a natural red-brown pigment – suspended in linseed oil. This concoction was called 'Falu redpaint': and many of the high-performing Scandinavian finishes still give at least a nod to the original Falu redpaint, in their late-20th and 21st century formulations.)

7.18 Non-film-forming finishes

The key thing about the Madison Formula and its descendants is that such finishes do not form a film or a skin on the surface of the timber. They tend to penetrate into the surface much more; and they also resist cracking (see Figure 7.5). Instead, they 'erode' gradually: and when they eventually do fail, by just fading away, it is simply a matter of refreshing the finish with one or more coats of new stain; because they don't require stripping off or sanding down, prior to redecorating.

Figure 7.5   Exterior stain finish showing some erosion – but no peeling or flaking

The really great thing about these 'breathable' finishes is that, because they do not create any kind of vapour-tight film on top of the wood, they will allow any retained surface moisture to evaporate through them: so that they do not break down in the same way that paints or varnishes do; nor do they trap water and thus create dangerously high levels of mc behind the surface layer in the decorated timber.

Wood stains are so named because of course they colour the wood to a greater or lesser extent; and yet they do not hide or obliterate the natural 'look' or figure of the timber surface. But be careful once again with names: don't confuse exterior wood stains with wood dyes.

These latter products are used to change the surface colour of *interior* wood, prior to varnishing or polishing it. Dyes always need another coat of something on top of them to seal the wood against the ingress of dirt: they are not the final finish in themselves.

7.19   Exterior paints

As I said a little while ago, these other finishes are essentially the same as wood stains and they work in the same way; but they differ in the fact that they obliterate the surface appearance of the wood (see Figure 7.6). They are still 'breathable' in that vitally important way, so they too protect the wood from the worst effects of excessive moisture content and consequent deterioration,

Figure 7.6   One of the newer generation of exterior paints

by not trapping water beneath them. And as with the stain finishes, their maintenance is much easier: just another coat or two on top, after a quick brushing down, with no sanding or scraping off to worry about.

7.20   The durability of exterior finishes

I have said that these finishes effectively 'erode' away as they age, rather than flake or peel off: but the evolution of the modern exterior wood stains has resulted in a great variety of possible surface appearances, ranging from pale, almost unpigmented, to very dark and highly pigmented: and from matt, to eggshell, to virtually full gloss. For ease of reference, these are generally known as 'low build', 'medium build' and 'high build' finishes, depending upon the amount of pigments and resin binders that they contain. And there is more or less a continuum of the tendency to either erode or to flake off, depending upon the level of 'build' – based on their solids content – within these finishes.

Essentially, the lower the build, the better the erosion characteristics and the higher the build, the nearer these finishes get to performing more like conventional paints or varnishes. But even so, the use of a high-build exterior finish is still a much better bet externally than is a normal 'gloss' paint or varnish. As a rough guide, three to five years before re-coating is usual for most microporous finishes, if you want to maintain its good colour and not risk the greying (weathering) of the wood substrate.

7.21   The effects of lighter or darker colours

It should be obvious perhaps, but some thought needs to be given to the effects of dark colours on exterior woodwork. This is because, as a purely physical phenomenon, darker pigments absorb more heat than do lighter colours (see Figures 7.7). Time for a short digression again …

The interesting fact that lighter or darker colours influence heat radiation was first tested and proven in the late 18th century by an enterprising Englishman named Benjamin Thompson – who later became Count Rumford – in Bavaria, where he was working at the time for his Highness the Elector of Bavaria, Crown Prince Carl Theodore, on improvements to the army. Thompson experimented a great deal with heat and thermometers (Bavaria gets very cold in the winter: down to minus 25 or 30° C) and so Count Rumford soon discovered that the traditional dark fur coats which everyone wore, lost more heat by radiation than did white fur. So he recommended wearing white in winter, rather than black: and they thought he was mad. But he was in fact quite right!

Anyway: it is a fact that heat is both radiated and absorbed more readily by dark, matt surfaces than it is by white, shiny surfaces. So on sunny days – even in winter – black or dark brown wooden surfaces will always be much warmer than lighter ones: and they will therefore tend to dry out much

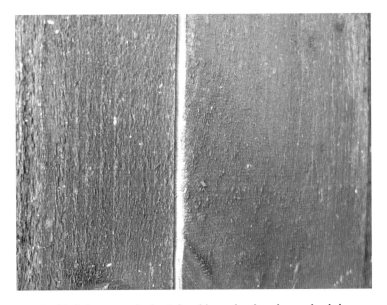

Figure 7.7   Shrinkage on dark-stained boards showing paler joins

more readily, and to stay relatively dry. (I myself have recorded mc levels as low as 12 % on dark-stained exterior cladding on South-facing walls in summer.)

For this reason, it is advisable to specify lower levels of moisture in exterior timber destined, say, for cladding or exterior joinery; but also to allow greater movement gaps or overlaps where these are required – such as with shiplap cladding profiles. Otherwise, boards can shrink more than expected and they will then expose pale, unstained edges: or worse still, they will 'pop' their joints and come apart – things that are most annoying to the unwary specifier or non-technical client, to say the least!

That's it on preservatives and finishes: now I need to give you a quick summary of what I've told you in this chapter.

7.22   Chapter summary

The first thing I made a point of saying was that wood does not normally absorb preservative throughout the entire cross-section of the timber component. So you will need to make sure that all 'working' of the timber – cutting off ends, notching, drilling large holes, etc – has been done *before* treatment.

If you *must* cut, notch or drill into any pre-treated timber on site, then please make sure that any of the cut or exposed areas of untreated timber are *re*-treated in situ; using a brush-applied concentrate preservative solution. (These products are readily available - usually from the same place where the original pre-treatment was carried out.)

I then discussed the basic methods by which wood preservatives are put into timber: high pressure and low pressure treatments. Then I looked at the types of preservatives available nowadays: harking back to 'CCA' – which is still available, if only in theory – and then going forwards to the modern formulations. Some of these still use Copper and others use combinations of rather fancy-sounding Organic chemicals like 'azoles' and 'permethrins', which act as either fungicides or (bat-friendly) insecticides. And I made the observation that none of the newer formulations have been tested in real life for the 60-plus years that CCA has shown itself to be so good at.

I also warned you that the term 'treated' is pretty meaningless on its own; and I advised you to specify treatment to one of the Use Classes that I gave you in Chapter 4.

Then I examined the field of Exterior Finishes for wood. And I started by warning you off the use of conventional paints and varnishes (especially 'Yacht Varnish!'); and I explained why the newer generation of so-called 'microporous' finishes are so much kinder to wood out of doors.

I commented on the weathering characteristics of exterior stains and 'proper' exterior paints and I tried to give you some idea of their longevity. I also noted the difference between lighter and darker colours and cautioned about potential problems with the latter.

So much for looking after wood in the higher-hazard situations. In the next couple of chapters, I want to look at some of the timbers that are commonly used in the UK. First of all, Softwoods.

8

Principal Softwoods
Used in the UK

Softwoods are – as I said in Chapter 2 – quite primitive in their cell structure: and as a consequence, there are fewer major differences between the different Genera and their species, than you will find amongst the Hardwoods. Of course, there *are* differences: I said right at the start that there is no such thing as simply 'wood' – but the point here is that the Softwoods have more in common with one another than do many of the Hardwoods, in terms of density, texture and so on.

In this chapter, I want to look in turn at each of the Softwoods that we use in the UK – or at least, those that we use in any sort of volume – and I plan to do that in a fairly logical way (well, that's the hope!). Therefore in this chapter I will be following – as indeed, I ought to – the basic parameters for the Naming of Timbers that I established in Chapter 2. So I will give you, first of all, the timber's Trade Name; followed by its one-and-only 'scientific name' – the name by which it *cannot* be confused with any similar-sounding wood species!

After that, I will discuss the timber's general appearance, its average density and its main characteristics. Then I will finish off each timber's section with a list of its normal uses – at least insofar as it tends to be employed here in the UK.

And I'll begin this chapter by describing the UK's two most commonly-used – and also most commonly available – Softwoods; which are known almost universally within the UK Timber Trade by their common names of 'Redwood' and 'Whitewood'. (Although, as you should know from my pedantic attitude all through this book: those are not, of course, their 'proper' names!)

8.1 European redwood (*Pinus sylvestris*)

This, as you will be able to tell from its scientific name, is most definitely a true Pine; and as I said, it is one of the two most commonly used Softwoods in the UK. It comes in a large range of qualities: from close-textured and almost clear, or with only very small knots; to relatively open textured, with large

Wood in Construction: How to Avoid Costly Mistakes, First Edition. Jim Coulson.
© 2012 John Wiley & Sons, Ltd. Published 2012 by John Wiley & Sons, Ltd.

knots and other defects such as wane and resin pockets. Each of the various qualities, as may be imported from many different Northern countries, is preferred – and priced accordingly – for different end uses: for example, joinery or construction.

Pinus sylvestris grows over a huge area: from just south of the Arctic Circle, in Norway, Sweden, Finland, Russia (and all of the former Soviet Union countries), the Baltic States; and also right through Central Europe, from Poland down to the Alps and even into Spain and Italy. Commercial production of European redwood, as imported into the UK comes principally from Sweden, Finland and Russia: with some occasional imports (usually the lower qualities) from Poland and Latvia.

Like all true Pines, the sapwood of European redwood is pale and creamy-coloured, whilst its heartwood has a more or less pinkish-brown tinge to it – and it's this reddish heartwood colour that gives the wood its Trade Name and common name. Like all the true Pines, it tends to darken on exposure to UV light; so that after some months, it takes on quite a pleasing, golden colour to the sapwood, whilst the heartwood darkens to a more rich red-brown.

The heartwood of European redwood is classed as Slightly Durable; so it needs to be treated with preservative in order to be used in exterior situations. But fortunately, it is rated as easy to treat with preservatives; and so its sapwood can be fully impregnated by the high pressure treatment process.

Trees from the more northerly, colder latitudes (Northern Scandinavia or Russia) are of much slower growth; and they produce timber of a much finer, harder texture than wood of the same species from elsewhere in Central or Southern Europe.

The average density of air-dried European redwood is about 500–520 kg/m³ although this varies considerably, depending upon its particular growth area.

When grown in the UK, this same species is properly called 'Scots pine'. It is one of our very few 'native' conifers (that is, it was not introduced to the UK from other parts of the world and then planted deliberately by foresters – although of course it is now re-planted commercially); and it's the only native species of Softwood from the UK that is used in any great commercial quantities.

The overall quality of British-Grown Scots Pine is not generally as good as that of imported Pine: and it is also not quite as strong a timber, in its natural state, as the imported variety. Therefore, the everyday Trade Name of 'Redwood' should be reserved *only* for imported timber from Europe, Scandinavia, Russia and the Baltic States. Commercial supplies of *Pinus sylvestris* from the UK should always be referred to as 'Scots Pine' – or more simply as just 'British Pine'.

The main end-uses for Redwood in this country are joinery, mouldings and PSE sections, construction (although not very much these days), wood panelling and exterior cladding, furniture, flooring, decking, fencing, packaging and transmission poles. British grown Scots Pine can be used for construction, fencing and packaging; but it is not used as a joinery timber.

8.2   European whitewood (mostly *Picea abies*)

This is the other main Softwood most commonly used in the UK: especially for structural uses. It is actually not a single wood species, but a combination of two unrelated European tree species, sold under a single Trade Name. These two different species are European Spruce (*Picea abies*) and Silver Fir (*Abies alba*).

Just as with European redwood, Whitewood is available in a range of qualities, each with its preferred end uses: although in the UK, we tend to go only for the lower qualities, for some reason.

Picea abies grows throughout the same geographical areas as *Pinus sylvestris*, whereas *Abies alba* is restricted more to Central Europe and the Alps. Therefore, commercial parcels of European whitewood from Scandinavia and the Baltic States will be pretty well 100 % Spruce; whereas whitewood from countries such as Germany, Austria and the Czech Republic will contain a moderate percentage of true Fir. In Switzerland especially, Fir trees grow to a very good diameter and so the timber is then processed and sold separately on its own, not as part of a Whitewood specification.

In appearance, European whitewood is bright and whitish or pale yellowish in colour, usually with no distinction between its pale heartwood and its pale sapwood (although sometimes the heartwood of Spruce may have a slight pale brown tinge to it). The pale colour of the wood is, of course, what gives the timber its Trade Name, and also its common name, of 'Whitewood'. Unlike Pine, even when exposed to UV light, it does not darken very much and so it can retain its lighter colouration for a long time.

The density of European whitewood is slightly lower on average than European redwood, being about 470–500 kg/m³, when seasoned.

Again, just as with European redwood, the Spruce trees that grow in colder Northern latitudes produce a very much finer-textured and better quality of European whitewood, whereas trees that grow in milder climates produce more open-textured and lower-quality timber. However, Alpine whitewood from Southern Germany and Austria is also of a very good, close texture: this is because it grows in the mountains, where altitude compensates for latitude, in terms of the effects of climate.

The heartwood of Spruce is generally reckoned to be of low durability, so it needs to be treated with preservative for any long-term exterior use: and it is not really suitable for very long exposure to ground contact, since – unfortunately – it is also rated as being Extremely Resistant to preservative treatment.

European Spruce as a tree species is not native to this country; so when grown in the UK, it is more correctly known as Norway spruce – or simply as 'British Spruce' (although it must be said that the individual species, *Picea abies*, makes up only a small part of all the so-called 'British Spruce' that is grown and harvested in the UK).

In the same way as with the relative qualities of Scots Pine versus imported Redwood, you will find that British-grown Spruce is much faster grown and of poorer overall quality than imported Whitewood. It is also not as inherently strong as imported timber of the same wood species: and this fact is most important when using Strength Graded timber (see Chapter 6).

The main uses of European whitewood are joinery (although much less commonly found in this use in the UK, than European redwood is), flooring, furniture, construction, fence rails and garden fence posts, packaging and pallets. At the moment, mainly because of price considerations, European whitewood is by the far the most widely-used Softwood for general construction in the UK. British Spruce is also suitable for all except the top quality end-uses, such as joinery and flooring.

8.3   Sitka spruce (*Picea sitchensis*)

Another true Spruce, this tree is actually a native of North America, although it is also one of the most commonly-planted timber species in the UK: forming by far the main component of British Spruce. But unfortunately, its lower density and lower bending strength make it less suitable for the higher-strength construction uses, than either imported redwood or whitewood – or even British grown Scots Pine.

Because Sitka spruce is of course a true Spruce, its heartwood is also of very low natural durability: and furthermore, as a true Spruce, it is also very resistant to preservative treatment. However, on the bright side, its movement characteristics are rated as Small.

As a timber, Sitka Spruce is very similar in appearance to European whitewood: and because it, too, is an introduced species, and also because it is grown in large plantations; these two factors mean that British-grown material is generally much coarser in texture, with larger and more numerous knots. It is also – like Norway Spruce – weaker than its imported equivalent.

The main uses of Sitka Spruce in this country are construction, fence rails, garden fence posts, packaging and pallets.

8.4   Western hemlock (*Tsuga heterophylla*)

This timber is not related to either the Pines or the Spruces (as you should be able to tell from its scientific name): although it does have a superficial resemblance to Spruce, and perhaps more so to a true Fir. (In Canada, Western Hemlock and some true Firs are frequently sold together as a Species Group, traded as 'Hem-Fir' – and I'll elaborate on the idea of Species Groups a bit later on in this chapter.)

Western Hemlock is principally imported from Western Canada, where the trees grow to extremely large diameters (1.8–2.4 m) and great heights (60 m or so).

Therefore the imported timber is available in long lengths and in large sizes, as well as in 'clear' (knot-free) grades, owing to the very great size of the tree. It also grows all the way down the West Coast of the USA into California, and Eastwards as far as the Western side of the Cascade Mountains.

Commonly referred to simply as 'Hemlock', it is a pale, creamy-white timber, sometimes showing a pinky or purplish sheen, and occasionally showing narrow, dark lines running along the grain. It is generally very slow-grown, which gives the timber a good density – around 490 kg/m^3 – and a pretty good surface hardness (for a Softwood!), making it ideal for high-class joinery and turned components, such as newel posts for stairs. (This last item is, at the moment, one of the most likely components in which to find Hemlock in the UK.)

In most other respects – such as its working properties, permeability, and so on – Hemlock is pretty similar to European whitewood. It also shows no great contrast between its pale brown heartwood and its whitish sapwood, indicating that it is of relatively low natural durability: in fact, it is rated only as Slightly Durable.

Although Hemlock has been planted to a small extent in the UK, it really doesn't play any great part in the harvested volume of British-grown Softwood production.

The uses of Hemlock in the UK are high-class joinery, external and internal doors, staircases, turnery, construction, decking and fence rails. However, up until the early 1990s, the relatively good price of Canadian Softwoods versus Scandinavian timbers at that time, meant that a huge volume of constructional timber imported into the UK was in fact Hemlock: and if you were to look at all the domestic floor joists in 1980s housing stock, you'd find they are mostly Hemlock.

8.5 'Douglas fir' (*Pseudotsuga menziesii*)

This tree is a native of Western North America, growing in pretty much the same areas as Western Hemlock; and being most abundant in British Columbia, Washington and Oregon. (In fact, its two main historical production areas are reflected in two of its 'alternative' – and of course, very old-fashioned – trading names: 'Columbian Pine' and 'Oregon Pine'.) Douglas Fir has also been planted to quite an extent in the UK: especially in Scotland.

This timber is *not* – despite its common name – a true Fir (nor, despite its two old-fashioned names that I just quoted earlier, is it anything to do with Pine either!). That is why its name, when used in any Standards or technical references, is always written 'in inverted commas' – to show that it is not a *true* representative of the wood type that it uses as its Trade Name. But, as you should know by now, its common name is liable to be misleading; especially if the Timber Trade should have anything to do with it!

Douglas Fir, like Hemlock, is a very large tree, producing timber in good, long lengths and large sizes, as well as being available in 'clear' grades. Even

British-grown material is available in large sizes; but as you might now expect, the timber grown in this country is not quite as good, in terms of either visual quality or strength, as that imported from the North American continent.

The density of Douglas Fir is typically around 530 kg/m^3 – which is a bit higher than European redwood: hence its reputation for slightly better strength. (It is at least as strong as European redwood or whitewood: although Canadian sources frequently claim that it is stronger than most other Softwoods.)

The wood itself is very distinctive: having a deep orange-red heartwood and also a very marked contrast between the earlywood and latewood parts of the growth ring, thus giving rise to a very pronounced 'flame' figure in flat-sawn material, and a strong parallel 'stripe' figure when it is quarter-sawn (and which the North Americans – wrongly, of course – call 'comb grain': because this figure is caused by the pattern of the rings, and *not* by the orientation of the grain). Its sapwood, by contrast, is pale and creamy-coloured: although commercial parcels of Douglas Fir are unlikely to feature much sapwood, since it is usually trimmed off at the sawmill, on account of the very large size of the tree; and thus there will be only a relatively tiny proportion of sapwood in any of the boards, compared with the much greater volume of heartwood.

The heartwood of Douglas Fir is rated as Moderately Durable – which means that it can be used for Use Class 3 without any additional preservative treatment. It is classed as being resistant to treatment with preservatives: but with extra processing, it can be successfully pressure-treated to levels of retention which will enhance its natural durability sufficiently well to enable it to be used satisfactorily in Use Classes 4 or 5.

The uses of Douglas Fir are marine timbers, vats and tanks (especially in breweries), joinery (especially for doors), construction, fencing: and also for plywood – where its good strength and natural durability are a distinct advantage. It is only occasionally imported into the UK these days, and generally only in two qualities or types: either as very large, heavy beam sections (300 mm × 300 mm and so on) or as so-called 'Door Stock' for high class joinery uses.

Before I leave Douglas Fir, I would like to tell you an unusual twist in the story of how it was identified. It is named after a Scottish plant collector of the Victorian era: David Douglas. He was collecting specimens in the newly-colonised Canadian province of British Columbia (of which the city of Victoria – on Vancouver Island – is the capital) and he needed to get as much plant material as possible from this enormous tree, in order to be able to describe it fully, for the purposes of giving it a scientific name. The general description of the tree was easy enough; but in order to properly describe its needles and cones (an essential part of its overall taxonomy) he needed a definitive example of a branch, bearing all the right stuff. Now, since Douglas Fir grows to over 80 m high – that is, over 250 feet – and its first branches are over 30 m – 100 feet – from the ground, you can see that it would be no easy matter to simply lop off a branch and take it home for analysis. So he *shot one off* with his rifle: now, that's what I call good, practical wood technology!

8.6   Larch (mainly *Larix decidua* and *L. kaempferi/L. leptolepis*)

Larch, unusually amongst the confiers, is *deciduous* – that is, it loses all its needles in winter.

It is not native to the UK: although in its natural state, European Larch (the first of the scientific names quoted above) is widely spread throughout all of mainland Europe. Japanese Larch (the second pairing of scientific names given above – which are alternatives) has also been planted in Britain, especially in Scotland and Wales.

All of the different species of Larch are quite similar in appearance: and in fact, they strongly resemble Douglas Fir in both colour and overall character, having a very marked growth ring figure, with a pinkish heartwood surrounded by a narrow band of creamy-coloured sapwood.

The average density of Larch grown in the UK is about 530–590 kg/m³, which gives it a very good structural strength. In fact, its strength is very similar to Douglas Fir, as is its natural durability: its heartwood being rated as Moderately Durable.

In the UK, commercial Larch is not usually grown to as large a diameter as Douglas Fir, so it is not often used for any significant construction uses. It also tends to be very knotty, and it is often prone to numerous black, dead knots, which tends limit its uses to packaging, fencing and the like.

However, some Larch is now being selected for exterior cladding (where its natural durability is an advantage); and these qualities have to be relatively knot-free; or with only sound, tight knots present in the boards.

Another species of Larch (*Larix occidentalis*) – known as Western Larch – grows naturally in British Columbia: and we can sometimes find it in the UK, where it may be included as part of a shipment of 'DFL' (see later in this chapter).

8.7   'Western red cedar' (*Thuja plicata*)

This is another of those *ginormous* North American West Coast tree species: growing to well over 70 m in height and around 2.4 m in diameter. It is slightly unusual for such a large and slow-grown tree, in that it is of a much lower density – only about 370 kg/m³ – than the other common Softwoods. Also, as may be inferred from its very low density, it is of much lower strength than all the other commercial Softwoods. For this reason, it is not used as a structural timber: although it can be used for simpler constructions such as conservatories, where its high natural durability rating is a distinct advantage. And as you should be able to tell by now, since I have put its name 'in quotation marks', it is *not* a true Cedar, despite its Trade Name. (True Cedars are in the Genus *Cedrus* – but of course, you expected something like that by now!)

The timber is very distinctive in colour: with a reddish-brown heartwood which can vary in its actual colour tone from a sort of salmon pink to a dark chocolate brown, when freshly-felled; but settling down, upon exposure to light, to a much more uniform russet-brown colour.

Quite unusually for a Softwood, the heartwood of Western Red Cedar is rated as Durable. It is classed as being resistant to preservative treatment; but – like Douglas Fir – it can take treatment well enough to enhance its natural durability rating to cope with more hazardous uses, or just to increase its potential lifespan in service.

The main uses of Western Red Cedar are greenhouses and conservatories, shingles and shakes (these are effectively, 'wooden tiles' – see the Glossary at the back of this book) and exterior cladding.

As a final point on this particular timber: you will need to be aware that Western Red Cedar is quite acidic; and it will thus accelerate the corrosion of metals in contact with it: especially iron. Because of this, either stainless, zinc-dipped or coated fixings must be used with it, for any outdoor or high humidity situation.

8.8   Southern pine (*Pinus spp* – principally *Pinus elliottii* and *P. palustris*)

This is the 'villain' of my story concerning confusing timber names, that I told you in Chapter 2: and it's a timber that is still being referred to in the UK Timber Trade as 'Southern Yellow Pine', despite that name being discouraged and its 'official' name (even in the USA!) being properly 'Southern Pine'. It is not – as you can see from the lack of a single scientific name above – one individual species of Pine: it is more properly a Species Group, consisting of about five or six closely related species, and being sold under a common Trade Name. It comes from the Southern States of the USA – principally Louisiana, Texas, South Carolina and Florida – with some variation in the character of the timber, depending upon which individual species is being considered.

There is a further confusion with this timber, since it was historically sold as 'American Pitch Pine'; and so it can often, in my experience, catch out carpenters and joiners, who mistake examples of its use in historic buildings and swear blind that they are dealing with true 'Pitch Pine', when this timber is really nothing of the sort. (Trust me: I'm a Wood Scientist!)

Southern Pine is a very dense and strong timber – for a Softwood: varying from 660–690 kg/m^3 depending upon which exact wood species it's made from. That means it is about 20 % heavier (and also stronger) than European redwood. As its (albeit incorrect) older, historic name suggests, it is very resinous ('pitch' being the American term for the word 'resin' that we normally use here in Europe). This fact makes it far less suitable for joinery work; where its strong tendency to resin exudation can give trouble with many surface finishes. (As an interesting aside: much of the world's production of natural Turpentine comes from the resins extracted from trees in this Species Group.)

Because of its good strength, Southern Pine is extremely suitable for building uses and for external structures such as timber decks.

Like all Pines, it has a reddish heartwood, surrounded by a pale sapwood; but unlike European redwood, its heartwood is rated as being Moderately Durable. Its sapwood, as with all true Pines, is rated as Permeable and is therefore easy to treat with preservatives. In surface appearance, it has a very strongly-marked growth ring character; with pale yellow earlywood bands alternating with much darker, more resinous latewood bands.

Its uses are principally construction – internal and external – and exterior cladding. It has been used for flooring and staircase building: but some care is needed when selecting it, to avoid those excessively resinous pieces.

8.9   Yellow pine (*Pinus strobus*)

This is the 'innocent party' from my Naming story in Chapter 2: since this one is officially allowed to be referred to as 'Yellow Pine'. As a timber, it is imported into the UK mainly from Eastern Canada – from Quebec and Ontario – but as a tree, it also grows down the Eastern seaboard of the USA, and even as far South as Kentucky.

Yellow Pine is indeed another 'true' Pine – as its scientific name *Pinus* shows you – and it therefore shares many of the characteristics of all the Pines. It has a pinkish heartwood (although this is quite pale, in comparison with most other Pines) and a pale creamy-coloured sapwood. Its growth rings are, however much more inconspicuous, with little or no earlywood and latewood contrast to them. And also unusually for a true Pine, its timber is not at all resinous: which makes it very good for domestic joinery work and even for cabinet-making. It is also unusual in that it has very low moisture movement characteristics: and this factor makes it a highly sought-after timber for making industrial patterns (these are the original timber 'moulds' which are used for precision metal castings); since it is dimensionally extremely stable under damp, factory conditions.

Although it is easy to treat with preservatives, this particular characteristic is almost never called into use, since the timber is almost always used in interior situations.

The main uses of Yellow Pine in the UK are for joinery and engineering pattern-making: although it can also be found in use as very large-section beams in old Mills and other industrial buildings, dating from the mid-to-late Victorian period.

8.10   'Parana pine' (*Araucaria angustifolia*)

This timber is one of the most unusual of the commercial Softwoods. Unlike all of the Softwoods that I've described so far – which have been Temperate in origin – this one is a Tropical tree species, being a native of South America. It grows in the high plateaux of Brazil and also in Paraguay and the Northern

Argentine. Of course, it is *not* a Pine at all – as both its scientific name and the fact that its Trade Name is written 'in inverted commas' should by now have told you! It is in fact closely related to the more well-known garden variety of 'monkey puzzle' tree.

It is a pale, creamy-coloured and very close-textured timber: which often shows bright red and/or dark greyish-blue 'mineral streaks' running along the grain. Its growth rings are indistinct, owing to the fact that it grows in the Tropics, and so it has no proper 'stop and start' to its growth cycle. It also grows as a very impressive tree, reaching up to 36 m in height and 1.2 m in diameter; and as a consequence, its timber is available in long lengths and almost knot-free qualities. Despite a good many Timber Trade 'old wives' tales' to the contrary, Parana Pine really *is* a Softwood; although I've personally come across many, many wood people over the years, who are convinced that it is a Hardwood. It's not.

On the down side: the timber has sometimes been reported as being quite prone to distortion on drying and (quite unusually for any timber) a tendency to shrink in its length. It is not a durable timber; although its sapwood is permeable: but neither of these attributes is of any great significance, since Parana Pine is almost never used in any high-hazard situations.

Its uses in the UK are primarily as staircases, joinery and occasionally wood panelling. It is becoming increasingly hard to find in the UK: not because it is in short supply commercially, but because there are some unresolved issues surrounding its Certification in respect of forest sustainability. But if you want it, you can find it: even in the 21st century.

8.11   Species groups

I promised a little earlier to tell you a bit about some of the Species Groups that are available: mostly from Canadian sources. These are generally group-ings of quite unrelated timbers, which are collected together and sold commer-cially for a number of end uses: and they are really not the same thing as, say, Southern Pine; where closely-related trees are combined into a single Trade Name.

8.12   Spruce-pine-fir

This of course is not a single wood species, but it is a very common North American species group: being made up of quite a high number of individual species of true Spruces, Pines and Firs (but *not* of course 'Douglas Fir' – which you know is not any sort of a 'proper' Fir). These trees all grow together in vast stands, throughout huge areas of the USA and Canada; and for ease of harvest-ing they are graded and marketed together under this one name – which is often abbreviated to 'SPF'.

Just occasionally – usually only in their native countries – individual species will be separated out and sold for specific end-uses (such as Lodgepole pine for joinery, for example); but normally these timbers will only be seen in the UK as part of the overall species mix, with Spruce predominating.

The principal species that make up SPF are Black spruce, Western White spruce, Engelmann spruce, Lodgepole pine, Jack pine and Alpine fir.

The main uses of SPF are principally in construction and in packaging. There may be some joinery uses from selected species, but currently this is quite rare: only because the price is presently too high, because of the Dollar exchange rate versus the Pound and the Euro. There is nothing at all wrong with the timber!

8.13   Hem-fir

This is another of those North American species groups, coming principally from the Western USA and British Columbia. It consists of Western Hemlock and Amabilis Fir (hence the name!); the two of which are grown, harvested and sold together as a single product: mainly in structural grades. Once again, it is rare to find it in the UK, owing to the UK's currency fluctuations making it currently too expensive, when compared with Baltic whitewood.

8.14   Douglas fir-larch

This is the final species group that I want to mention here. It consists of Western Larch (*Larix occidentalis*) mixed in with a proportion of true 'Douglas Fir': both of which are native to the Pacific Coast of the USA or Canada. More usually known as 'DFL', it is principally sold in the structural grades; but once again, there is not much volume of it in the UK at present, owing to its relatively high price, compared with UK alternatives.

Well that's it for Softwoods. But this time, I don't think you need me to give you a Chapter Summary for this one: if you need to know anything about a particular timber – just look back a few pages from this point! (I recommend familiarising yourself with these timbers and their characteristics: it will help you a great deal, if you ever need to specify something for a particular job.)

9 A Selection of Hardwoods Used in the UK

You may have noticed, after digesting the preceding chapter, that the majority of the Softwoods are quite similar to one another: both in their properties – such as density – and in the majority of their possible uses. But the same thing cannot be said about the Hardwoods.

Whereas the Softwoods, due to their more primitive and very much more basic cell structure, are quite restricted in the ways in which they can show themselves, the Hardwoods can – and do – show a huge range of colour, texture, figure, density and strength that is almost limitless in its variety.

This is because, as I told you in Chapter 2, the Hardwoods are much more highly evolved and they have a much greater number of cell types than do the Softwoods: and they also have many more different ways of rearranging those specialised cells. Amazing as it may seem, there are quite literally tens of thousands of species of Hardwoods in the world: although in purely commercial terms, only a few dozen different ones are ever imported regularly into the UK in any quantity.

There is not the space here to give you an exhaustive list of all the Hardwoods that are available commercially in some shape or form; in fact, there's not even the space to describe those which could potentially be found in the UK if you scoured every Importer's or Merchant's yard. So instead, I will concentrate on those timbers that are more regularly seen here – and many of which have been in regular use in this country for decades. Even though they are perhaps 'familiar' timbers, it is still important to know a bit more about them, so that they may be used correctly. After all, you know the old saying: 'familiarity breeds contempt'.

But first, just to give you a 'taster' of the exciting and rather more exotic world of Hardwoods, I'm going to mention just a few timbers which – whilst not on my larger list to be described in detail in this chapter – may excite your imagination, by the sheer magic of their names.

Afzelia; Banga Wanga (yes, really!); Chickrassy; Danta; Eng; Freijo; Gedu Nohor; Hyedua; Ironbark (no prizes for guessing what that's like!); Jarrah; Karri (both of these last two are species of *Eucalyptus*); Louro; Makoré; Nyatoh; Owewe

Wood in Construction: How to Avoid Costly Mistakes, First Edition. Jim Coulson.
© 2012 John Wiley & Sons, Ltd. Published 2012 by John Wiley & Sons, Ltd.

(you really need to say it out loud); Padauk; Quaruba; Rauli; Sterculia; Thitka (not to be confused with 'Sitka' – unless you have a lisp! – oh, and its alternative name is 'Kashit' … sorry); Ulmo; Virola; Wallaba; (no 'X'!); Yang and, finally, Zebra Wood (yes, it *is* stripy!). What a fantastic set of words! They're also fantastic timbers: and I have samples of 90 % of them in my office. (I admit it: I'm a wood anorak.)

Now – just as with the Softwoods – I will first of all give you the Trade Name of each timber that I intend to describe in detail. I'll give its Common Name and any (sometimes confusing) variants to that name: and, of course *always* its Scientific name. I will then describe the timber and point out any particular features or characteristics that could make it suitable for its various or usual end uses.

If possible, you should try to obtain for yourself a sample of any timber that you're interested in using. But be warned: a single sample cannot hope to show you the great variety and range of colour and figure that the actual timber can exhibit when used in a large amount, such as in a run of panelling, say, or a shop fit-out.

In the previous chapter, I presented the Softwoods more or less in the order in which you would be likely to encounter them: with the most commonly-available and most used described first. But with the Hardwoods – since no individual species tends to predominate, and because each individual species is used in much smaller volumes than are any of the Softwoods – I have simply presented them in Alphabetical Order.

9.1   Ash, American (*Fraxinus spp*)

This timber – as the name might suggest! – comes from North America: but it is only found in the Eastern half of the USA and Canada. I've listed it as 'Fraxinus spp' because American ash (and remember – you've come across this notion before) is really just a Trade Name: meaning that the timber may consist of more than one actual species of wood, albeit that they are closely related.

Sometimes the different species of *Fraxinus* may be sold separately: in which case, they are given names which all relate to colours. *Fraxinus americana* is sold as 'white ash'; *F. pennsylvanica* as 'green ash'; and *F. nigra* as 'black ash' or sometimes as 'brown ash'. The most likely species to find being sold as a separate parcel of timber is the last one: black ash.

American ash trees can reach about 30 m in height and up to 0.9 m in diameter. The timber is generally light in colour (almost white in *americana*, though a bit darker in *nigra*), with the pale sapwood not clearly demarcated from the heartwood: although the latter can sometimes have a grey-brown or reddish tinge to it. Its grain is very straight and – since it is a ring-porous timber – it has very obvious growth rings and a coarse texture.

In terms of its density, ash can weigh about 660 kg/m³ when seasoned (black ash is a generally bit lighter in weight, being about 560 kg/m³) and its heartwood is rated as only Slightly Durable; although it is easy to treat with preservatives. However, it is rare that anyone would bother to give ash any preservative treatment, since its predominant uses are for furniture and internal joinery.

As a timber, its outstanding property is its toughness – a property which gives it very good shock resistance. (No, I don't mean that you can't electrocute it!) You can knock it about – such as in a hammer or pickaxe handle – and it will be very resistant to cracking or breakage. By and large, American ash is used for most of the things for which European ash is used.

9.2   Ash, European (*Fraxinus excelsior*)

This is a close relative of American ash, and it grows all across Europe and into Asia Minor. Trees can reach a height of about 30 m and a diameter of up to 1.5 m: so it can produce larger boards than its American counterpart, if need be. In appearance, it is very similar to American white ash, although it may be pinkish when freshly cut, though this soon fades. Very occasionally, the heartwood may be brown or black: this is not due to any rot – in fact, its cause is not really fully known – and such timbers can command a higher price, especially when cut into veneers. European ash is a bit denser than American ash, averaging about 690 kg/m³ when seasoned.

European ash also has good toughness properties – the best of any British-grown timber – and it's easy to work, on account of its straight grain. It is used for furniture, joinery and shopfitting; but also for tool handles and – almost uniquely – for the rungs of wooden ladders: also on account of its toughness. There is one slight quirk, however, in relation to slow or fast-grown material: where faster-grown timber is preferred for both tool handles and ladder rungs, because it is considerably stronger (see Chapter 2 for an explanation as to why this is so).

9.3   Beech, European (*Fagus sylvatica*)

There are a number of different species of beech in the temperate parts of the world; but the only commercially significant one, in UK usage terms, is the European variety: which grows throughout most of Northern and Central Europe and western Asia. There is some British-grown beech used in the UK, but a large proportion of the commercial timber used here comes from France, Germany, Denmark and Rumania: and much of this European material is usually steamed, which then gives it much more of a pinkish tinge.

In its natural state, beech is a pale, light-brown timber, with no differentiation in colour between heartwood and sapwood. It is classed as Not Durable, although it is easy to treat with preservatives: but as with ash, it is unlikely to be employed for many outdoor uses (except perhaps for the odd fence post from a small home-grown sawmill).

Beech trees can grow to 30 m in height and about 1.2 m in diameter. The density of beech is about 720 kg/m³ when seasoned and it is a very strong timber, with good steam-bending properties; which makes it an ideal choice for furniture and also for plywood (bentwood) manufacture. In its cell structure, it is a diffuse-porous timber, with a very fine and even texture, which takes paints, stains and adhesives well. These factors are also strongly influential in making it a favourite for furniture making: though it is also used for tool handles, brush backs, toys and turnery.

However, there is one property of beech which must be treated with caution: it has very large movement characteristics. So for any use where major or seasonal changes in atmospheric humidity are anticipated, you really must take great care with the detailing, in any design where overlaps, butt joints or tongues and grooves are a feature. Beech flooring – although popular – can be problematic, if its equilibrium moisture content (see Chapter 3) is not carefully considered before a final design and methods of installation and fixing are chosen.

9.4   Birch, European (mainly *Betula pubescens*)

I love the botanical name of this! (Its common name is 'hairy birch' – but it conjures up more than that, I think?) Birch grows throughout a large part of Europe, but it is most common in northern and eastern Europe; especially Scandinavia, Russia and the Baltics: where its main use is for the manufacture of plywood. In fact, birch is little used in its solid form in the UK and it's almost never seen except as plywood; so I'm going to refer to it only in the chapter on wood-based panel products.

9.5   Cherry, American (*Prunus serotina*)

This is one of the very good, decorative Hardwoods available from the eastern USA. It is an attractive timber, having a pale sapwood and a pinkish-brown heartwood that contains darker lines and often shows a greenish tinge. It is diffuse-porous, with a very fine texture, which makes it extremely valued for attractive furniture and panelling.

Cherry produces only relatively small trees – up to about 20 m or so in height and about 0.6 m in diameter – and its density is around 600 kg/m³ when seasoned. Because of its small tree size, cherry is generally only available in narrower boards; and its uses are confined to cabinet making and certain specialist uses, such as bridges on stringed instruments. However, it is also produced in veneer form, which means that its attractive figure can be seen and used in shopfitting, for example, when veneers are bonded to a suitable substrate such as MDF.

9.6   Chestnut, Sweet (*Castanea sativa*)

This timber species is the wood of the edible chestnut that you can roast on the fire (the nut, that is, not the wood!).

Sweet chestnut grows throughout central and southern Europe and it has been planted in Britain for centuries: it was once very popular as a coppice-grown timber, for producing poles. The tree can grow up to 30 m in height and about 1.5 m in diameter: and its timber closely resembles European oak (but without the very large rays that are a particular feature of oak). It has a pale sapwood and a golden-brown heartwood. It is ring-porous and therefore coarse in texture; but straight grained and very strong. It is less dense than oak, being about 540 kg/m³ when seasoned; with good natural durability and small movement characteristics.

The timber of Sweet chestnut has a variety of uses, but principally these are furniture, gates and fencing; and in this last use category, it is often seen in the form of cleft (i.e. naturally split) staves, wired together to form a picket fence. It has also now been strength-tested – just in the past 20 years or so: so there are now figures available for it, in the four Temperate Hardwood Strength Grades (see Chapter 6 for the details of these).

9.7   Ekki (*Lophira alata*)

This is a tropical timber, grown in several West African countries. Ekki is the name we use for it in the UK; but in Francophone countries, it is called 'azobé' (just to be difficult!). It grows as a very large tree, up to 55 m in height and 1.8 m in diameter; with a pale sapwood and a very dark red, or even chocolate-brown, heartwood. It is an extremely heavy timber, with a density somewhere in the region of 1000 kg/m³ – occasionally more! Unsurprisingly, it is very strong, but it is also very difficult to work with, not least because of its highly interlocked grain: and it is rated as Very Durable. The main uses for ekki are lock gates and piles for jetties or docks. It has also been used to some extent in the UK for outdoor decking.

9.8   Greenheart (*Ocotea rodiaei*)

This is another very hard, extremely heavy, tropical timber: but this one comes from northern South America; Guyana and Honduras. Trees can grow to 40 m in height and up to about a metre in diameter. Its density is an amazing 1030 kg/m³: and you won't be surprised to learn that it is the hardest and heaviest of the timbers that are commercially available today. In appearance, it is dark olive-green (hence the name!), with a pale green or yellow sapwood. Its heartwood is – of course! – Very Durable.

Greenheart is almost exclusively a structural timber: it's used mostly for heavy engineering jobs; especially in marine environments, because of its outstanding resistance to marine borers. But it does have one other, rather surprising, use: as very high high-class (and highly expensive) fishing rods – because of its extremely high bending strength and good modulus of elasticity.

9.9   Idigbo (*Terminalia ivorensis*)

Idigbo is a tropical timber, coming to us principally from Nigeria and Ghana. And, as with other West African timbers, the Francophone nations have another name for it: 'emeri'. It is a very tall tree, reaching a height of 45 m, with a diameter of up to 1.2 m. The wood is yellow or light yellowish-brown, with a slightly paler sapwood. It can have a slight tendency to interlocked grain, but normally, it is quite straight grained: with a medium-coarse texture. It has strong growth rings (quite unusual in a tropical hardwood) and when flat-sawn, it has somewhat of the appearance of oak: and it is quite often used as a substitute for oak, especially by shopfitters. This may well be on account of its movement characteristic, which is rated as Small.

Idigbo is extremely variable in its density: ranging from only about 370 up to as much as 740 kg/m^3 when seasoned: which again is a very unusual characteristic in a tropical hardwood. And, like oak, it is rated as Durable: so it needs no preservative (that is, if it should ever be used in situations where that property might be needed, which I have to say is unlikely). Another characteristic that it shares with oak is the tendency to corrode iron fixings or fittings when wet: and it also contains a yellow ingredient, which can bleed out in wet situations – so it is not recommended for kitchen equipment (especially not chopping boards or draining boards).

The primary uses of idigbo in the UK are for furniture and high-class joinery; although it could actually be used for exterior cladding, if care is taken to minimise the effects of the yellow dye-stuff washing out and staining a concrete render, or something similar.

9.10   Iroko (*Milicia excelsa*)

This is another tropical timber, whose growth range extends into East as well as West Africa. There are some other 'native' names for this wood, but iroko seems to be the one 'fixed' name that all of Europe uses to trade with (thank goodness!).

It is a very large tree, reaching about 50 m in height and 2.5 m in diameter. It is quite heavy, averaging 640 kg/m^3 when seasoned, and it has small movement, which makes it very suitable as a joinery timber – especially when coupled with its rating of Very Durable: meaning it can be used out of doors without any preservatives. The wood colour is yellowish-brown to brown, with a pale sapwood. It has an irregular grain structure, often interlocked, with a coarse texture. Another very useful attribute is good acid-resistance.

The main uses for iroko are joinery (both interior and exterior), flooring and laboratory benches: but it is also rated as a structural timber (making Strength Class D40 when properly graded).

9.11   Keruing (*Dipterocarpus spp.*)

This timber comes from a wide area of the Far East: growing in Burma, India, Thailand, Malaysia and the Philippines. It is sometimes given the local name from its different places of origin, such as 'gurjun', 'yang', 'apitong' or 'eng' – although keruing is the name used for exports to the UK.

Keruing is a plain brown timber, without any sign of an attractive figure, and its trees can grow up to 60 m in height and 1.8 m in diameter. The wood is commonly red-brown to dark brown in colour, and it is moderately coarse in texture, though with almost no tendency to show any interlocked grain. Its density is about 750–800 kg/m³ when seasoned. It is rated as only Moderately Durable, and it has medium to large movement characteristics. Keruing is also well known for its tendency to exude large amounts of gum: which means that it is not particularly suitable for joinery uses.

The uses of keruing in the UK are somewhat limited: largely due to its high gum exudation. Its main uses are for the thresholds in pre-assembled door casings, light industrial flooring, and lorry or trailer floors. It is also rated for structural use: although I have to say, it is seldom used for trusses or beams, in my experience.

9.12   Mahogany, African (*Khaya ivorensis and K. anthotheca*)

This timber is found throughout West Africa; and it consists mostly of the first species, *Khaya ivorensis*. As a tree, it grows up to 60 m in height and 1.8 m in diameter; with a density of around 700 kg/m³ when seasoned. The timber is pink when freshly sawn, fading to red-brown: and its sapwood is creamy-white or yellowish. It often has interlocked grain and it has a moderately coarse texture. Its movement characteristics are described as Small and its heartwood is rated as being Moderately Durable.

African mahogany is employed in the UK almost exclusively for joinery and shopfitting uses.

9.13   Mahogany, American (*Swietenia macrophylla*)

American Mahogany comes from Northern South America and it is really the 'proper' mahogany: being closely related to the very original 'Spanish' or 'Cuban' mahogany from the West Indies, which is *Swietenia mahagoni*. As a tree, it grows to about 30 m high and about 1.8 m in diameter. It is an attractive wood; being light red-brown, with a high lustre: and it is seldom found with much interlocked grain – though in fact, it frequently shows some other very attractive variations in grain, including roe, curl, mottle and blister figures. Its density when seasoned is about 540 kg/m³ and it is rated as durable, with its movement characteristics being rated as Small.

The uses of American mahogany are high class joinery, furniture, cabinet work and engineering pattern-making (this last being the same as for Yellow Pine, if you remember).

9.14   Maple (*Acer saccharum*)

Also known in the UK as Hard Maple or Rock Maple, this is a very hard-wearing and tough timber, which comes from eastern Canada and the northern and eastern states of the USA. The timber is creamy-white, with occasionally a reddish tinge and its sapwood is also creamy-white. Its density is about 720 kg/m^3 and its natural durability rating is Moderately Durable; and it has Medium movement characteristics. (Of course, its other well-known attribute is its ability to exude a sugary sap – check out its scientific name! – which the Canadians and the more northerly Americans, such as those in the state of Maine, boil down to make Maple Syrup. It's great on pancakes!)

Because of its excellent wearing characteristics, Rock Maple is a very good flooring timber, especially for dance floors and other public areas, such as squash courts and bowling alleys. It is also used for furniture and panelling, as well as in sports goods.

9.15   Meranti (*Shorea spp.*)

This is not one single type of timber, but a group of over 20 timbers of different species, all from the same Genus, *Shorea*. Meranti grows in a very wide geographical area all over south-east Asia; principally in Malaysia. Similar timber of the same Genus, but grown in the Philippines, is known as Lauan. Commercial supplies of meranti are divided into two broad categories, based notionally on colour and density, and these are sold as Light Red Meranti and Dark Red Meranti.

Light Red Meranti is reckoned to vary in density from 400–640 kg/m^3, whereas Dark Red Meranti is supposed to be from 580–770 kg/m^3: so that means there is a mid-point, around the 600 kg/m^3 mark, where any piece of timber could be allocated into either one group or the other! The natural durability of meranti varies from only Slightly Durable in the Light Red variety, to Durable in the Dark Red variety: so it is really quite important to be sure what you are getting. (For this reason, it is recommended that meranti should be treated with preservatives when used for external joinery; since we can't be sure of its exact durability when left untreated.)

Meranti has been used for a number of years in the UK as a substitute for true mahogany: but it is quite prone to interlocked grain and it doesn't have any of the attractive variations in figure that true mahogany can show. However, it is a very popular (cheap!) timber for joinery – both internal and external – and also for shopfitting. Many of the numerous species of *Shorea* can also be found in the large volumes of cheap far-eastern plywood that come to the UK each year.

9.16   Oak, American red (principally *Quercus rubra* and *Q. falcata*)

This is another of the temperate Hardwoods of the eastern USA. Despite its name, the timber may not be very red: although the heartwood can sometimes show a reddish tinge. Its density is about 770 kg/m³ when seasoned: which is slightly higher than both American white oak and European oak, although it is reckoned to be not as good as white oak for decorative uses.

Unusually for oak, American red oak is rated as only Slightly Durable in its heartwood, which means that it is unsuitable for exterior joinery. It is also unusual in respect of another basic wood property: it is completely porous – which makes it totally useless for barrel making – so the whiskey distillers of the Southern States had better make sure they buy the right sort of oak to mature their famous products!

The uses of American red oak are principally for interior joinery and furniture or cabinet making.

9.17   Oak, American white (principally *Quercus alba*, *Q. prinus*, *Q. lyrata* and *Q. michauxii*)

As you can see, American white oak is not one sort of oak, but a mixture of at least four species: so once again, it is a Trade Name for a 'species group'. Despite its name, the timber is not 'white' but pale golden brown in the heartwood, with a white sapwood. However, it can sometimes show a pinkish tinge to the heartwood: therefore wood colour is *not* a reliable indicator of the tree type.

Its density is about 750 kg/m³ and it is a very straight-grained, although coarse-textured wood: and since it is ring-porous, it shows a very prominent growth ring figure on flat sawn surfaces. Like all true oaks, the timber has very deep and broad rays, giving rise to a highly attractive 'silver' ray figure on quarter-sawn surfaces: which is often made use of in decorative veneers.

American white oak – unlike its red cousin – is 'tight' and so it is used extensively for cooperage by the whiskey distillers that I mentioned a few moments ago. Its heartwood is rated as Durable: and it is also a moderately strong timber. In fact it has – only within the past few years – been tested and approved for use in the UK and Europe to the Temperate Hardwood Strength Classes (see Chapter 6).

9.18   Oak, European (mainly *Quercus robur*)

This is the oak that we are most familiar with in the UK. It grows all across Europe and we buy timber from many sources; so it tends to take the name of those sources as its description: hence French oak, Danish oak, Rumanian oak, and so on. 'English oak', as the name implies, is from our own forests; and it tends to be generally less straight-grained than its imported equivalent: but then it has perhaps a more interesting character as a result.

European oak is slightly less dense than the American oaks; being about 720 kg/m³ when seasoned, and it is correspondingly not quite so strong. Its heartwood is yellowish brown, and it has a wide, light-coloured sapwood: and this must of course be removed if the timber is to be used out-of-doors without preservative treatment (since only the heartwood is rated as Durable).

The uses of oak cover all the usual possibilities: from furniture making, to joinery – both internal and external – to construction. Its one potentially serious drawback, which needs some thought and care when using it in certain situations, is that it is a very acidic timber: which means it will severely corrode any unprotected iron fixings that may be used with it under damp conditions, especially in outdoor situations.

European oak – and more especially English oak – has been used in this country for centuries: for furniture, joinery and construction, as I've said. Now, maybe I'm alone in this: but I can't help feeling that architects and designers often specify oak out of a sense of 'tradition' or 'heritage': but if you were to examine its range of properties and compare them with what you needed for any specific job, then it really isn't always the best timber that you could use for many of the things for which it actually does get used. In particular, its long-term performance out of doors – in respect of weathering and surface splitting – is quite poor (because of its very large rays, which are a 'weak spot' in its grain structure): so in my view, a much better choice would be Sweet Chestnut, which has all the decorative features of oak, but without those enormous rays.

9.19   Obeche (*Triplochiton scleroxylon*)

This is another West African timber and – like a most woods from that part of the world – it has many alternative names, the most common of which is 'Wawa'. It grows as a very large tree; up to 55 m in height and 1.5 m in diameter: but despite that fact, it is a very lightweight timber, being only about 380 kg/m³ when seasoned. It is a very pale-coloured timber, with no clear distinction between its heartwood and sapwood: and of course (as you know by now) this means that it has no great natural durability, being rated as only Slightly Durable. However, it has a Small movement rating, which makes it very suitable for internal uses where its stability is an asset.

Obeche has quite strongly interlocked grain and a coarse texture: so it tends to be used where appearance is not important: for example, in the framing of upholstered furniture. From my own experience, I have seen obeche used extensively for the bench seats in saunas: where its low density gives it excellent insulation properties – therefore being cool to sit on – and its lack of any resin or gum exudation is also a great advantage – after all, you don't want to go sitting on a blob of hot resin with your bare bottom, do you!

9.20 Opepe (*Nauclea diderrichii*)

This is yet another West African timber that has been in use in the UK for well over 50 years. It is a very large tree, growing up to 50 m in height and 1.5 m in diameter. Its sapwood is pale, but it has a very distinctive heartwood, that is often bright yellow when freshly-felled; but fading to darker orange on exposure to UV light. It is rated as Very Durable: although because of its coarse texture and a tendency to have interlocked grain, as well as its really high density – up to 750 kg/m³ – it is used almost exclusively for heavy engineering applications, such as jetty piles and wharf timbers, as well as for lock gates and railway sleepers. Opepe has also been tried as a flooring timber; so of course, it can be used for decking, where its durability rating means it would need no preservative.

9.21 Sapele (*Entandrophragma cylindricum*)

This is probably the one hardwood (with the possible addition of oak) that everyone will have come across, at some time in their working lives – and quite probably without realising it. That's because it is almost universally used, in veneer form, for office desks and flush doors. It's the 'stripy hardwood' that you'll see everywhere, in buildings designed or fitted out from the 1970s to about the turn of the new Millenium. (At the moment, there is a fashion for the paler timbers, such as maple or beech: but fashions being what they are, that will no doubt change again, and sapele will make a comeback sooner or later. Meanwhile, there's still plenty of it to be seen!)

Sapele is another of those timbers from West Africa, which have been popular for a very long time with UK hardwood users. As its scientific name implies, it has a very cylindrical trunk, with very little taper, and it can grow up to 60 m high: so each tree produces a lot of very high-quality timber, especially veneers. Its heartwood is a pleasing red-brown colour, with a very marked interlocked grain, which when quarter-sawn, gives it a very strong 'stripe' figure that is its trademark, so to speak. Its density varies from 560–690 kg/m³ and it is reasonably strong: although it is never used for structural purposes.

Sapele has been used for furniture and flooring; but its primary use, as I said earlier, is for veneers; and it has been universally employed for office furniture and for veneered flush doors. (If you look around your own office, or those of your local Council or college, you're bound to come across it somewhere, sooner or later.) In fact, sapele veneer is so abundant that it is frequently found as the 'balancer' (that is, the veneer glued onto the back face, which evens up the board's construction and helps to limit distortion) on decorative-faced panels of plywood or MDF.

9.22   Tatajuba (*Bagassa guianensis*)

This timber comes from South America and it is one of the newer timbers to come onto the UK market (although it has been known about for at least 40 years!). It is related to iroko – which of course comes from West Africa – and it is fairly similar in both its appearance and its uses, although it is somewhat heavier, averaging around 830 kg/m³ when seasoned. Its heartwood is orange-brown when freshly-sawn, darkening on exposure to light, and its texture is quite coarse.

Tatajuba is rated as Very Durable, so it is a very good timber for outdoor uses; and indeed, its primary use in the UK to date has been for decking.

9.23   Teak (*Tectona grandis*)

This is a very well-known timber, which has been used in the UK for many years; and indeed, it was highly popular in the 1960s and 70s as a furniture wood – and now it seems to be making a bit of a fashion comeback, especially as garden furniture.

Although native to Thailand, Java and (especially) Burma, we don't see much of the indigenous timber here in the UK. Most of what we see here now has come from one of the many extensive plantations around the tropics – particularly from Africa and – increasingly – South America.

It is an attractive, golden brown timber, with dark streaks and an attractive figure, and with a moderately coarse texture. Its density is about 650 kg/m³ and its grain is usually very straight.

It is rated as Very Durable and it also has Small movement characteristics; both of which properties make it ideal for external joinery and garden furniture. Its chief characteristic feature is its natural oil, which migrates onto its machined surfaces quite rapidly, thus giving it a 'greasy' feel – but of course, this helps it to remain relatively weatherproof.

Apart from external joinery and furniture, its other well-known use is, of course, for the decks of very expensive boats!

9.24   Utile (*Entandrophragma utile*)

Look at the scientific name and you'll soon see that utile is closely related to sapele – and it comes from the same part of the world. (Oh, by the way, I must emphasise that it is pronounced 'You-Tilly' and *not* 'Ewe-Tile' – as though it should be on a roof! All West African timbers whose names end in 'e' have that final letter stressed. Pedantic moment over.)

Utile is an attractive red-brown timber, with some tendency to have interlocked grain – though not usually so strongly as sapele. And perhaps for this reason, it is more often seen in the solid, rather than as a veneer. Utile is a little heavier than sapele, averaging about 670 kg/m³ and it is just slightly coarser in texture. Its heartwood is rated as Durable.

Utile is sometimes used for furniture, but more often for interior and exterior joinery – and for the latter, its natural durability rating is a great asset of course.

9.25   Walnut, American (*Juglans nigra*)

The scientific name for the species rather gives away its other common name of 'black walnut'; and indeed, the UK Timber Trade often call it, in full, 'American Black Walnut'. It forms a moderate-sized tree, about 30 m high and up to 1.8 m in diameter; and it grows on the eastern side of the USA and a little way up into the eastern part of Canada. As a timber, it is quite hard and dense, being about 640 kg/m³ when seasoned and its heartwood is very dark brown, deepening with age.

The very pale sapwood is clearly demarcated from the very dark and decorative heartwood; but in commercial shipments of this timber, the sapwood is not usually excluded from the graded boards (it is sold in a quality that is often described as 'sap no defect' – where 'sap' of course really means 'sapwood').

Its deep-coloured heartwood does, in fact, reliably indicate that it is Very Durable: although American walnut is another of those timbers whose high price and highly specialist uses, mean that this particular property of the timber is unlikely to be tested very much. Its main uses are for very high quality furniture and the stocks of extraordinarily expensive shotguns.

9.26   Walnut, European (*Juglans regia*)

You can see from the name of the Genus that the European and American timbers are close relatives and they have a similarity in texture and character: although the English and French walnuts are not so dark in colour as the American sort. In terms of both density and strength, the European timber is much the same; but its heartwood is only rated as Moderately Durable (though again, it is unlikely to be exposed to much of a hardship). Once more, its preferred uses are furniture (often in veneer form) and gun stocks – and speaking of stocks (pardon the pun), not much comes from England these days; most supplies of European walnut come to the UK from France or Italy.

9.27   Whitewood, American, or Tulipwood (*Liriodendron tulipifera*)

This is one of those timbers with many different names – it is even called 'Yellow Poplar' although as you may tell from its scientific name, it is not at all related to that wood. And to call it 'whitewood' as well … (See European Spruce, if you've forgotten the link.)

Tulipwood – to use its less confusing appellation – is a good, all-purpose furniture and joinery timber, that is wonderfully easy to work with and which takes stains and glues extremely well.

As a tree, it grows in the eastern USA, where it gets up to 30 m high and 2.5 m wide. Its density is quite moderate – only about 500 kg/m^3 – and its heartwood is a yellowish or olive-brown, whilst its sapwood is very white. It is rated as Slightly Durable.

The wood is very soft and is easily worked, so it is very popular for interior joinery and shopfitting; and because it takes stains so well, you may often come across it disguised as another timber – and therefore not recognise it in disguise, so to speak.

Well there you are: a selection of Hardwoods as used in the UK at the moment. And, as with the Softwoods, that's about it for this chapter, as you won't need me to give you a Chapter Summary for this one either. If you need to know anything about a particular commonly-used Hardwood, then it's easy to check back a few pages. And, as with the Softwoods – although possibly even more so – I strongly recommend that you familiarise yourself with these timbers and their characteristics, in order to help yourself, if you ever need to specify one of them for a particular job.

And if you need any information about some timber that I haven't covered here, then I've put a list of helpful organisations and publications in the reference section at the back of this book.

10 Wood-based Sheet Materials

In this chapter, I want to examine the specific features of the various different sheet materials that are in use in the construction industry (in which category I also include joinery and shopfitting uses as well as just 'building'). In doing this, I want to point out the ways in which the individual make-up of the different types of wood-based panels can affect the jobs that they can, or can't – or sometimes shouldn't! – be used for. These sheet materials are all made out of wood, in some way shape or form (so I don't intend to talk about plasterboard or any other type of non-wood panels); and I will divide these sheet materials into three basic 'families'. This is partly to enable me to cover them in some sort of logical order, but also because the manner of their construction really does affect the usable properties of each of them in quite a significantly different way: since these three basic families reflect the fact that the resultant sheet materials come from using wood in ever-more divided or subdivided ways.

Plywoods all use slices or 'plies' of wood (which are, of course, veneers). Particleboards (which category includes chipboard and OSB) all use either small chips or thin 'wafers' of wood. And the Fibreboards – as their name so obviously implies – all use the fundamental elements of wood, which are the fibres themselves.

10.1 Plywood

I'll now make a start on this whole topic by explaining how plywood is made, and what the different types are that you might find on the market. But be warned at the outset: there are very few 'proper' standards which are applied to plywood around the world. There *are* in fact a number of European Standards covering plywood (there are really quite a lot of them: though many of them are of no huge significance or importance to the final end-users of plywood. But some of the European Standards decidedly *are* quite important to know about). However, you will find that the majority of the huge volumes of plywood that are imported into, and bought and sold in the UK do not comply with most of the Standards that exist.

Wood in Construction: How to Avoid Costly Mistakes, First Edition. Jim Coulson.
© 2012 John Wiley & Sons, Ltd. Published 2012 by John Wiley & Sons, Ltd.

10.2   The two fundamental properties of plywood

The first thing to recognise about plywood is that its constituent parts – its veneers – are usually glued together in a 'cross-banded' construction (or 'layup'). By cross-banded, I mean that alternating veneers are turned at 90 degrees to one another: so that each veneer then has the one immediately above or below it rotated by a quarter turn. This results in a panel layup whereby the grain direction of each alternating veneer runs at right-angles to its neighbour above and below, right throughout the thickness of the plywood panel. (There are some speciality and 'engineering' plywoods where the veneers may sometimes run in different – or even the same – directions: but the plywoods that are used in any form of building or joinery uses all have this most common cross-banded layup.)

And this cross banded construction confers two very significant properties on plywood. First of all, it helps to more nearly even out the tendency for the whole sheet of plywood to 'move' across the grain. That's because, as each alternate ply tries to swell or shrink *across* its width, it is at the same time being restrained by the ply immediately above or below it, which does not want to move *along* its grain (remember, wood does not swell or shrink appreciably along the grain): and so the plywood panel as a whole is much more 'stable' than solid wood can be, in response to any moisture content changes.

The second property in which plywood differs from solid timber is in its overall strength characteristics. You will know, from what I said in the early chapters of this book, that the strength of timber lies mostly along the grain: with its cross-grain strength, by comparison, being in the order of 40 times weaker. So by cross-banding the grain orientation in alternate directions, plywood is thus able to even out this inherent disparity: and so making it into a material that can be loaded in any direction, without there being any particularly 'weak' direction to the panel. That is why plywood has been so successful in use as a 'sheathing' material. (This term means that it can cope with diagonal stresses as well as with stresses along or across the panel: and so it can be used as a load-bearing diaphragm, when fixed to something like a timber frame panel or as part of a stressed-skin floor construction.)

Oh – one more thing about the cross banded layup of plywood: there will almost always (with a very few, very unusual exceptions) be an *odd* number of veneers in each sheet (see Figure 10.1). This is because the requirement to alternate the grain orientation would result in the direction of the outer or 'face' veneers being in opposing directions – that is, at right angles to one another – if there were an even number of veneers (just count it out yourself and see). Doing so would then result in the panel being out of balance; and so that would make it very much more liable to distortion under changing moisture conditions. (For this reason, when manufacturers bond an additional decorative face veneer to any plywood – or indeed, to any other wood-based sheet material – they will always bond a cheap 'balancer' veneer

Figure 10.1 Exposed veneer edges – showing cross-banded construction of plywood

onto the back face as well; with the grain of this balancer veneer oriented in the same direction as the decorative face veneer.)

10.3 Basic types of plywood

All 'normal' or 'standard' plywoods will have a layup based on the cross-banded method that I've described above. But there are lots of different types of plywoods on the world market: and they can be divided up in a number of different ways. But to my mind, the most logical way is to look at the basic wood material that the different sorts of plywood are made from. This way of describing them will also – as you'll see – result in the users of plywood being able to identify some other important characteristics about them: things which will follow naturally from their basic wood ingredients and their particular and detailed layup.

The different sorts of plywood can be readily separated into Hardwood and Softwood plywoods, of course. But more than that; they can be further separated into more generally recognisable subdivisions – so I'm now going to talk about the following generic types: Conifer plywoods, Temperate Hardwood plywoods and Tropical Hardwood plywoods: and each of these types has particular characteristics which make them somewhat different from each of the others.

10.4 Conifer plywoods

As the name suggests, these are made up from veneers of various coniferous (i.e. Softwood) species: generally Spruce from Scandinavia; whereas North American plywoods will use a mixture of a few different Softwoods (Douglas Fir, Spruce, Pine and occasionally Hemlock).

Figure 10.2   Birch Plywood (left) and conifer plywood (right) showing greater or lesser number of veneers in the same thickness of panel

The main factor which marks out all Conifer plywoods is that they will generally consist of an odd number of veneers, of a constant and uniform thickness: a veneer thickness which is actually quite substantial in ply terms – around 2–3 mm. And, apart from those plywoods which are available in a 'sanded' face finish (where the sanding operation has thus reduced the thickness of the face veneer by a small amount), every veneer is the same thickness throughout the panel. Conifer plywoods therefore tend to have fewer veneers than some other plywoods – such as Birch plywood, for example – even though their overall panel thickness may be the same (see Figure 10.2).

Conifer plywoods are the ones most commonly used for 'engineered' or 'designed' structural purposes: and in the main, they have been properly tested. Therefore the Timber Design Codes (which was, latterly, BS 5268: Part 2 in the UK – but which has just been withdrawn at the end of 2010 – and now also for today's timber engineers, Eurocode 5) will list a small number of Conifer plywoods coming from different sources – essentially, countries such as USA, Canada or Finland, which can be used by structural engineers with confidence, to design and create stressed skin panels, timber frame house walls, and so on.

10.5   Temperate hardwood plywoods

The only representative of this type which you are likely to see used in construction – and even then, not very commonly – is Birch plywood. And the 'Rolls Royce' of these is Finnish Birch, which has an enviable reputation for quality and reliability; but it is very highly priced (and highly prized!) and so only a few dedicated timber engineers tend to use it, to create things like box beams or other diaphragm-type structural components. Birch plywood is also made in Russia and Latvia, and its manufactured quality can vary quite a bit: so my advice is to look for some sort of Quality Assurance Certification (ISO 9001or similar) as part of its 'provenance' before you commit to using it to actually build with.

If you look at Figure 10.2, you'll see that Birch plywood has many more, very thin veneers to it, as compared with Conifer plywood. This is partly because the very fine texture of Birch allows it to be sliced into much thinner veneers: but it is also because the greater number of veneers will give us a panel much of more evenly-distributed strength. (A Conifer plywood which has seven veneers will have three of them running in one direction and four of them running in the other; thus making a 3:4 strength ratio – whereas a Birch plywood with 13 veneers will have six running one way and seven running the other; thus making a 6:7 strength ratio. And that is much closer to being 1:1 – so the Birch plywood will be much more 'equal' in strength properties, either along or across the panel, than will the Conifer plywood.

Birch plywood has been popular for many years, for use as a good 'paint grade' sheet material; because of its fine texture and its generally 'clear' surface appearance: although nowadays, it has been somewhat overtaken by MDF. Of course for non-structural uses, the quality of its manufacture is far less critical; but even so it can be very annoying (to say the least) if the veneers should fall apart (we call this 'delamination') or if blisters or bumps should appear on the surface when the plywood gets damp – or even whilst it is being painted or stained. Therefore, a good, trustworthy manufacturer and a reliable source of supply are highly important. And all of that is generally what is wrong with the third type of plywood.

10.6   Tropical hardwood plywoods

These are the most common type of plywood seen in the UK; and they are the most varied – and indeed, variable in quality. There are a large number of manufactured types: these days coming mainly from Malaysia and Brazil; but much more frequently in the past, coming from Indonesia (although question marks over Indonesia's forest practices have largely curtailed the trade in this new Millennium). And then there is China. Although not strictly a 'tropical' country, its plywood manufacturing practices are nevertheless

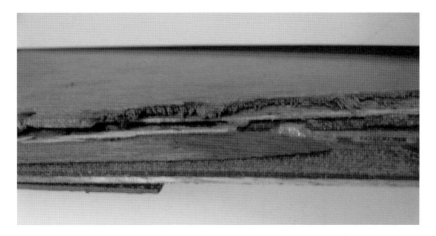

Figure 10.3   Delaminated veneers in a tropical hardwood plywood

pretty well identical to those of the countries I have just cited: and plywood from China bears a marked similarity, in its layup, to all of the others within this group.

Tropical Hardwood plywoods all have these features in common: they are made from a motley assortment of mixed species – often of varying degrees of Natural Durability; they are often very suspect in their glue-lines (delamination is a not infrequent occurrence); and their layup consists of one or more very thick 'core' (interior) veneers, with extremely thin front and back face veneers, normally less than 1 mm in thickness.

Why, then are they so popular? Two reasons: they are cheap, and they look nice – that is, they generally have very high quality appearance 'best face' veneers, which give them an instant appeal to those who know no better; and for whom beauty is (literally) only skin deep. But their huge drawback is that they generally – with only a very few exceptions – have little or no Quality Control in their manufacture and they are therefore not in any way guaranteed to perform well in any permanent building use. The major difficulty with these plywoods, as I have frequently verified from my many years of problem-solving with wood-based products, is delamination, caused by poor or inadequate gluing (see Figure 10.3).

10.7   Plywood glue bond and 'WBP'

Let me say – categorically, and with no possibility of being misquoted or misunderstood – *THERE IS NO SUCH THING AS WBP PLYWOOD!*

Those three initials have, collectively, done more harm for the reputation of plywood, and wasted more construction money, than any other single factor you could name. If I had a pound for every instance where I've come across some dubious plywood, described – sometimes on the material itself, but always somewhere on the paperwork associated with it – as 'WBP' then I wouldn't need to worry about any royalties from this book!

The initials WBP originally stood for 'Weather and Boil-Proof' and they were meant to be a sort of 'guarantee' that the plywood was well bonded: it was supposed not to fall apart under extreme moisture conditions. But over the years (and it is quite a few decades now), this initially accurate description has been watered down – almost literally, sometimes – until it means nothing much at all, other than a vague idea that the plywood might have some sort of 'quality' to it. But I am telling you here and now – *do not trust it*! The initials WBP certainly nowadays mean that what you are getting is a cheap and nasty plywood that cannot be trusted to do its job properly or reliably.

But not only that: the term WBP only ever existed in the now-obsolete British Standards for plywood, that have long since been replaced by the many European Standards that I referred to earlier. So WBP has been a 'non-term' for over 20 years … though you can still buy it from third-world producers, who are prepared to give you what you say you want; even if it may be completely invalid and obsolete!

10.8   Exterior

'All right then,' you may say: 'if WBP isn't the correct term and plywood labelled as such can't be trusted, then what should I be asking for instead?' And my answer to that is one word: 'Exterior'.

This term is a very specific and proper term, which is given in one of those numerous European Standards that cover plywood: and it is given in EN 314 – or rather, Part 2 of EN 314, to be more precise. In EN 314-2: 1993 there are three separate classifications given for the glue-bond quality of wood based sheet materials; and which I reckon are relatively easy to remember, since they more or less describe the situations in which such materials may end up (or 'hazard levels', if you will).

These different degrees of the risk of exposure to moisture (which could result in delamination of the sheet material) are called:

(1)   Class 1 – Dry Conditions.
(2)   Class 2 – Humid Conditions.
(3)   Class 3 – Exterior.

Now what could be simpler? So in future, when you need a plywood that is supposed to perform properly in a situation where it will be exposed to prolonged wetting, or outdoor exposure, without it delaminating, you will know what to specify or use: Class 3 Exterior to BS EN 314-2 – and *not* some obsolete stuff that may be described (or even labelled) as 'WBP'. Here endeth the lesson. Almost.

10.9   Adhesives used in plywood

There are only two or three basic types of adhesive resins used in the manufacture of plywood: and the third type is not normally used alone, as you'll see. The first main sort of man-made polymer adhesive is Urea Formaldehyde – known as 'UF' for short – and the second main type is Phenol Formaldehyde – known either as 'PF' or, more usually as 'Phenolic resin'. The first sort is only slightly resistant to wetting and should really be used for Class 1 (Dry) uses; whereas the second one is completely water-resistant and can meet Class 3 (Exterior) uses without much problem. The third type of resin is Melamine Formaldehyde – known of course as 'MF – which has much greater moisture resistance than UF, although it is still not completely waterproof and nowhere near as good a PF for that property. MF resins tend to be very expensive; and so to reduce the cost, they are often blended with UF resins to produce the sort of glue bond that will satisfy Class 2 (Humid) uses.

And now here's a useful bit of advice on how to identify the glue bond that has been used in the plywood you are being offered. Not with 100 % certainty; but with enough certainty to know that you might not have a 100 % water-resistant adhesive. I'll explain.

Both UF and MF resins are transparent: so they are, in other words, colourless. But PF resins are not: they are dark red-brown and show up, very distinctly. So if you are offered something that has been glued and which is described as being a 'waterproof' (or even let's say, 'WBP') plywood, but you cannot see the glue-lines within it: then it *cannot* be an Exterior type of plywood – since it must have been manufactured with a transparent adhesive, which can only at best be a Class 2 glue bond and not Class 3. Easy peasy!

However, the reverse situation is not necessarily true. Just because you *can* see the darker-coloured glue-lines that does not automatically make it an Exterior plywood. Simply using the correct type of adhesive does not guarantee that it now meets all the requirements of BS EN 314-2. The factors that I described earlier – poor or inadequate gluing, generally resulting from a lack of good Quality Control at the factory – can easily mean that despite using the right materials, the result is still a rubbish product.

Take Marine Plywood, as a case in point. From the amount of 'marine ply' sold by the Timber Trade, you could be forgiven for thinking that there must be one heck of a lot of wooden boats being built every year in the UK. I'm sure there are a few of them made each year; but nothing like the numbers that would be needed to justify such a steady demand for a material whose primary attribute is supposed to be its ability to survive in the sea! Of course, most of the demand for 'marine ply' is actually from the building trade, who have tended to see it as a more reliable form of plywood for more hazardous uses – such as bathrooms or shower cubicles – where they anticipate that an enhanced resistance to moisture would be an advantage.

But in the majority of cases, the so-called 'marine' type of plywood that builders use is just as prone to delamination as any other sort of plywood that they are prepared to pay a paltry sum of money for. And that is because most of it is not 'proper' Marine Plywood: in other words, it does not conform to the appropriate British Standard, which is BS 1088.

10.10   BS 1088 marine plywood

The *only* plywood that can correctly be referred to as 'marine' is that which is manufactured in accordance with BS 1088-1: 2003. (Part 1of BS 1088 specifies the two types of Marine Plywood available: Standard and Lightweight. Part 2 of BS 1088 defines a method of testing, to prove whether or not the glue-bond is adequate.)

This British Standard imposes two vitally important requirements on any plywood that claims compliance with it. First of all is the fact that it must be manufactured from wood veneers which have a durability rating of Moderately Durable or better; and secondly, the Standard requires that its glue bond must be rated as 'Exterior' in accordance with BS EN 314-2 (which BS 1088-1 references in its appendices). All of this means that someone somewhere is supposed to have had the plywood tested, to prove that it fully complies with all of the BS 1088 requirements. But also it means that the factory where it was made must have a robust Quality Control system in place, to be able to prove that each and every batch of plywood that is manufactured is still fully in compliance with all of the requirements of BS 1088: and therefore it will not rot away or delaminate when in service.

Anything else which simply bears the epithet 'marine ply' but which is not proven or guaranteed in some independently-tested way, is therefore not fully in accordance with BS 1088 and should therefore simply not be trusted as being 'marine plywood'.

10.11   Plywood face quality

So much for the quality of the glue bond that you will need to specify and achieve, if any plywood is to perform reliably. But what should your plywood look like? And how can you be sure that it is the right 'quality' that you have specified, or need for the job you have in mind?

Once again, the Tropical Hardwood Plywoods are the ones that lead the field (if that's the right expression) in terms of kidology: by calling themselves names such as 'BB/BB or 'C/CC' and so on. But what do all those letters mean? And can you take those terms at face value (sorry about the pun)? No you can't. Because they have no definite rules by which they can be judged.

10.12   Appearance grading of face veneers

The most reliable rules for the grading of face veneers used in plywood manufacture are those which apply either to Conifer plywoods or to Temperate Hardwood plywoods: because, as I have just stated above, there are no published rules governing the face descriptions of Tropical Hardwood plywoods. All of the published rules work on the basis that any defects which affect appearance, and which are essentially the same as those used in grading solid timber (principally knots, when it comes to Softwoods) are permitted to be present, to an ever greater extent, the further down the list of applicable grades one goes.

10.13   Conifer plywood appearance grades

In North America, Coniferous plywood face veneer grades are designated by letter, from A down to D. However, there is no 'A' grade available nowadays (if there ever was), and so the 'nicest' looking plywood will be described as being 'BB'. This means that it will have a B grade veneer as the front – or best – face and it will also have a B grade veneer as the back – or worst – face: so it is in fact good-looking on both sides. In the UK, the best-looking Conifer plywood you are likely to see coming in from the USA would be a BC: which would then have a B grade best face and a C grade back face – though more usually (if at all) you would be offered a CD quality. This would in all probability be described more fully as 'CDX' – where the letter 'X' after the face veneer grade descriptions refers to the nature of the glue bond: and that is (of course!) Exterior.

In addition to the 'regular' lettered grades, there is an intermediate quality that fits in between B and C, known as 'C Plugged'. This face veneer has wooden (or sometime plastic resin) 'plugs' or patches in the places where large knots have been cut out of the veneer prior to the plywod's layup, in order to improve its overall appearance (see Figure 10.4).

Canada uses essentially the same veneer grade designations of B to D to grade the veneers themselves, but the plywoods from there are given further titles which more reflect their overall fitness for different uses. Thus a plywood with a B best face and a C or D back face will be referred to as 'Good one side' (always abbreviated to G1S); whilst a plywood with two D faces will be described as a 'Sheathing' plywood: this being a structural plywood, where appearance is secondary to strength. Plywood that has an improved layup (that is, one which uses more whole-sheet veneers in its core) plus a C quality best face is called 'Select Sheathing': and as well as looking just a tiny bit better – which it doesn't usually need to – it will have improved strength properties, which is most important.

In Europe, all Conifer plywood veneers are graded – as you might expect – in accordance with a European Standard, EN 635-3: its full title being 'Plywood – Classification by surface appearance – Softwoods'.

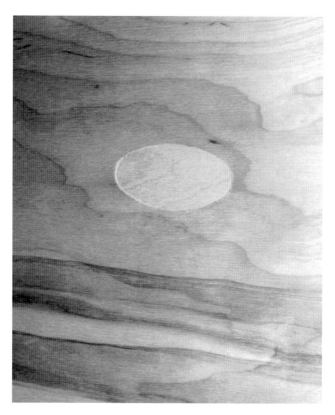

Figure 10.4   Plug in an improved face veneer

This Standard gives five appearance classes, but they are not, as you might have expected, the letters A to D with an extra 'improved' grade stuck in between. Instead, they use mostly Roman Numerals – except for the top grade, which in fact *is* a letter. This 'best' grade looks almost perfect; thus it is more or less impossible to find it being used in any quantity in Conifer veneers.

The BS EN 635-3 Classifications of veneer grades for Softwood plywood are known as E, I, II, III and IV. The two plywood qualities most usually seen in the UK are II/III and III/III: and these are almost always made in Finland, from Spruce. In pure appearance terms, these two types will be very similar to either a 'C Plugged C' or a 'CC' plywood from North America (and with the latter of course being known as a Sheathing plywood if it comes from Canada).

These types of Spruce plywood will always be a 'proper' Exterior quality, in terms of their glue bond. But please remember that Spruce is not a durable species of wood; so the plywood itself will need to have some sort of preservative treatment if it is to be used where moisture is a real risk. In other words, 'Exterior' as a term on its own *does not* provide immunity from decay. For that, you need a preservative – or a Marine Plywood (of the proper sort!).

These European Conifer (Spruce) plywoods are used in construction and shopfitting, by the more discerning users (or maybe by those who have had problems in the past with Tropical Hardwood plywoods!). That's because they are very reliable and they are guaranteed to perform, without delamination, and they don't vary overmuch in appearance from one sheet to another in a batch.

10.14   Temperate hardwood plywood appearance grades

European Temperate Hardwood plywoods should – in theory – follow the designations that are given in EN 635-3: 'Plywood – Classification by surface appearance – Hardwoods'. But in reality, the only reliable member of this plywood type is Finnish Birch; and the Finns have long had their own appearance classifications, which once again are based on letters: though it's not at all straightforward.

The best available Birch face veneer, in reality, is a 'B' grade (since 'A' grade is not really available). There is *no* 'C' grade: but instead, after 'B' comes 'S' and then 'BB' grade. Finally, there is another, called 'WG' (which I am told stands for 'well glued' – but that has never been confirmed by anyone with inside knowledge of the Finnish grading rules). However, you really don't need to bother with all of that; since the commonest types of Birch plywood that you will see in the UK are either B/B, B/BB or BB/BB. In other words, you may see any combination of two out of the usual three 'best' (or at least, 'not bad') appearance grades.

Before I leave the subject of Birch plywood, I ought to just warn you that there is a lot of 'other' Birch ply around: but it is not all from Finland. Most of the cheaper Birch ply comes from either Russia or the Baltic States – principally Latvia – and (as you ought to expect by now) it is not always very reliable. I have seen some good Latvian Birch plywood, but I've also seen some bad stuff as well: so my advice to you is to look for some sort of Quality Certification, such as that provided by ISO 9001. This at least shows that the factory takes its manufacturing responsibilities seriously. And of course, you need to check that they can assure you of its glue-bond credentials: that is, Class 3 Exterior, to EN 314-2.

10.15   Tropical hardwood plywood appearance grades

The more usual face descriptions that are seen in the UK in respect of Birch plywood have been borrowed (or rather, 'hijacked' might be a better word!) by the producers of Tropical Hardwood plywoods. But then those descriptions have been added to and further extended, beyond their original designations, until they have become more or less meaningless. So, as highly accurate descriptions of plywood, they certainly cannot be trusted: since there are no written-down rules which define what a 'BB' or a 'C' - or even a 'CC' – or any other supposed face grade of a Tropical Hardwood plywood should look like.

So please be aware that there are two things about Tropical Hardwood ply-woods that you should be pretty cagey about, from now on. First of all, you simply cannot trust its glue-line – especially if it says it's 'WBP' – but also, even if it claims to be 'Exterior' (after all, which Standard are they referring to? And has it been proven by any tests? Probably not!). And secondly, you won't know, from those meaningless and highly unofficial descriptions – such as 'BB/CC' – what its actual quality will be like until you see it. Which will be after you've bought it: and by which time it will be too late to complain about it.

So: have I put you off using cheap, nasty, un-Certified and untested ply-woods, based purely on their very low price? I hope so. There is an awful lot of cheap plywood on the market (or should that be, 'a lot of cheap, awful ply-wood'?). Beware, because most of it is highly unreliable.

If you want good plywood then you need to pay the market price for good quality – which is the same philosophy as anything else in this life. So unless you don't mind what happens to the job you're working on, I recommend that you use North American or Scandinavian plywoods that not only *claim* to meet 'proper' Standards, but which *do* – and which are also independently tested and Certified, to show that they really are what they say they are. And they also cover other matters of real importance, such as low Formaldehyde emissions, and other stuff you should worry about.

As a final reassurance, you should look for an independent Quality Stamp – such as CanPly (The Canadian Plywood Quality Assurance body) or APA (The Engineered Wood Association of the USA) – which will be your guarantee of independently-tested quality and reliability (see Figures 10.5a and 10.5b). Or you should use Finnish Birch plywood. (I'm sorry to say that the FinPly

Figure 10.5a   CanPly stamp

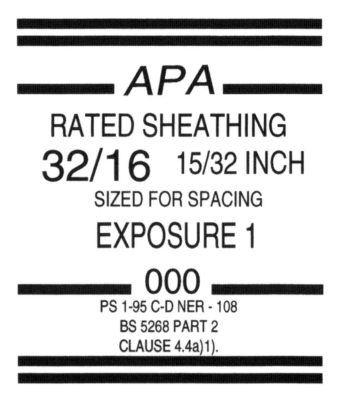

Figure 10.5b   APA stamp

stamp is no longer used, since there is now only one major producer of plywood in Finland. But their production quality is eminently reliable, I can assure you.)

But I repeat: if you are wanting a constructional plywood that will perform well every time; then please avoid Tropical Hardwood plywoods. That is, unless you can see guaranteed evidence of proper Quality Certification (and I can tell you, it's pretty rare).

Now I need to tell you a bit about Particleboards: though there's not really a huge amount to say.

10.16   Particleboards and wood chipboard

The main type of Particleboard that is used the world over, is wood chipboard. And as the name sort of implies, chipboard is *always* made from wood chips. That's not quite as daft a statement as it might sound, because there are a couple of other members of the particleboard family which look as though they are made of chippings, but they are not made from wood: although at first sight, they might look as though they are.

10.17   Flaxboard and bagasse board

These two members of the particleboard family are really only worth a brief mention; just to demonstrate to you that not all particleboards are made from wood. Flaxboard, as its name states, is made from the residue of flax plant stems (called 'shives') which themselves result from the manufacture of linen cloth. Mind you, here in the UK you're not very likely to see flaxboard 'in the raw', but it is highly likely that you may well be inside a building where that particular sort of particleboard has been employed: since its major use is as the internal core material of numerous fire doors. So you won't necessarily see it; but flaxboard will still be doing a valuable job.

As for Bagasse: that is the squeezed-out residue from sugar cane. I personally have never seen bagasse board imported or used in the UK; but as a board product, it is quite common in certain parts of the world – notably in South America.

But now to tell you a bit more about chipboard. Because of the fact that chipboard is a variety of particleboard – albeit the most common one – it is therefore governed by the European Standard for particleboard, which is EN 312: 2003. Because of this, all of the different chipboard types are designated by the prefix 'P'. In the UK, the construction industry is really only concerned with a very limited number of types of chipboard; nearly all of which are load-bearing, some of which are moisture resistant, and most of which seem to be used for flooring.

Particleboards use the same classification for exposure to moisture, as do the plywoods that are manufactured to European requirements. However, the adhesives used in chipboard are only suitable for use in Dry or Humid conditions; so the best category that you will get is 'moisture resistant' chipboard, since there is no Exterior quality available for any type of particleboard. All of the different chipboard types available under BS EN 312 are listed in Table 10.1.

Happily, the situation with regard to the record of quality and reliability with chipboard is a lot better than it is with plywood: since almost all of the chipboard which is imported into the UK is made either in Europe (in fact, a lot is even made here in the UK itself) or in countries that work closely to the European Standards. Therefore, if you want to specify or use a 'P5' chipboard for example, you should be able to find it quite easily: coming from a reliable and trustworthy source, correctly labelled and stating that it meets all the appropriate EN Standards. And the same thing holds true for OSB.

10.18   OSB

OSB is in fact simply another form of particleboard, and it was previously included within the former British Standard for particleboards: although nowadays it has a separate European Standard of its own, EN 300: 2006. The initials

Table 10.1   Table of particleboard types

Particleboard types given in EN 312: 2003	
EN type	*Use description*
P1	General purpose – dry conditions
P2	Boards for interior fitments (including furniture)
P3	Non-load bearing – humid conditions
P4	Load bearing – dry conditions
P5	Load bearing – humid conditions
P6	Heavy duty load bearing – dry conditions
P7	Heavy duty load bearing – humid conditions

stand for 'Oriented Strand Board' – though nobody ever actually refers to it by its full name. The board itself is manufactured from long, narrow strips (the 'strands' of its name), sometimes called 'wafers': and these are made from scraps or off cuts of plywood veneers.

The manufacturing process of OSB results in these strands being laid in a rough sort of 'cross-banded' construction, quite similar to the layup of 3-ply plywood (see Figure 10.6). This has the effect of making OSB a much more 'structural' type of board than chipboard, with very good strength properties both along and across the panel. Because of this, and also because it is typically somewhat cheaper than the good, properly-constructed 'sheathing' types of plywood, OSB has now been used in many of the instances where plywood was formerly employed.

Once again, the different types of OSB are designated by a series of letter-codes, which denote the types that are suitable for particular use categories (see Table 10.2). And also, as with chipboard, there is no fully Exterior grade of OSB available.

As I mentioned a moment or two ago, OSB has in recent years taken over a great deal from sheathing plywood, for many of the usual load-bearing situations: and most notably in the area of Timber Frame Housing. But if you are designing or constructing with OSB, you should be aware that it has a much greater tendency to expand (or 'move') – both along and across the panel – than plywood does. That's because its 'strand' type of layup means that it is only very approximately cross-banded: and since it is not made from large, more or less continuous sheets of veneer, the three 'layers' within the panel do not have anything like the same degree of restraint for one another that plywood has.

Therefore, you'll find that you need to allow somewhat greater expansion gaps between adjacent OSB panels, than you will with the same design using

Figure 10.6   Typical OSB surface appearance

Table 10.2   Table of OSB types

Oriented Strand Board types given in EN 300	
EN type	*Use description*
OSB/1	General purpose and boards for interior fitments including furniture – dry conditions
OSB/2	Load bearing – dry conditions
OSB/3	Load bearing – humid conditions
OSB/4	Heavy duty load bearing – humid conditions

plywood. In such cases, may I recommend that you use a 'conditioned' panel: this is one which has been allowed to absorb some moisture after its manufacture, and so it has already expanded somewhat. This will help to minimise any future problems related to moisture uptake in service.

Instances where over-expanded panels then butt tightly up against one another, can thus result in those panels bowing – or 'bellying' – outwards quite severely: and in the case of Timber Frame House walls, they have been known to block up the external cavity and so cause damp penetration and worse. So by all means use OSB instead of plywood: but please use it with some care and thought in your designs.

10.19   Fibreboards

This is the last of my three 'families' of wood-based sheet materials. They are all made, as indicated earlier, from that ultimate element of wood – its fibres. And there are in fact two distinct divisions within the family that is fibreboards: and these divisions are differentiated by their manufacturing processes, which are 'Wet' or 'Dry'.

10.20   Hardboard, medium board and softboard

These three are all variants of the Wet Process and they differ mainly in respect of their individual density range. As you might expect, Hardboard is the most dense and Softboard is the least dense, with Mediumboard occupying the middle ground, so to speak.

These board types are not overly renowned for their strength properties (except maybe for a few specialised designs that may use hardboard as part of a composite beam, for example). In fact, these boards are not much used in the UK these days: with the notable exceptions of 'Tempered Hardboard'and Softboard. This latter is still being universally referred to here in the UK as 'Insulation Board', despite that term having been discontinued by the European Standard which now governs all of the Fibreboards.

That governing Standard is EN 622: and in its five different Parts it covers every one of the different members of the Fibreboard family, including both Wet and Dry Process boards: as you can see if you refer to Table 10.3. (You may also spot that Tempered Hardboard is the only member of the fibreboard family that is rated as being fully Exterior.)

The various Parts of BS EN 622 have been revised and re-issued over the past few years. Part 1 deals with general requirements and was revised in 2003. Part 2 deals with Hardboards and Part 3 deals with Mediumboards; and both of these Parts were revised in 2004. Part 4 deals with Softboards and Part 5 deals with MDF; and both of these Parts were revised as recently as 2009.

Hardboard in the UK is used mainly for overlaying old floors, to give a more even surface – but if you are thinking of doing that sort of thing with it, please remember to 'condition' all the panels first before fixing them down. This 'conditioning' process very simply involves brushing water into the back face (or the 'mesh' face, as it is also known) and then allowing the panels to expand for a few hours; then fixing them down before they shrink again, so that they remain tight and flat.

Some quantities of hardboard and mediumboard are also used by the furniture and shopfitting industries, where a strong, thin and flexible panel is needed. And you may occasionally come across mediumboard being used as a sheathing panel, since it does have adequate strength for that purpose: but it is not used for that sort of thing in the UK, so you're only likely to see it as part of

Table 10.3 Table of fibreboard types

Wood fibreboards given in BS EN 622: Parts 2,3,4 and 5	
EN type	*Use description*
Hardboard	
HB	General purpose – dry conditions
HB.H	General purpose – humid conditions
HB.E	General purpose – exterior conditions
HB	Load bearing – dry conditions
HB.HLA1	Load bearing – humid conditions
HB.HLA2	Heavy duty load bearing – humid conditions
Low density mediumboard	
MBL	General purpose – dry conditions
MBL.H	General purpose – humid conditions
MBL.E	General purpose – exterior conditions
High density mediumboard	
MBH	General purpose – dry conditions
MBH.H	General purpose – humid conditions
MBH.E	General purpose – exterior conditions
MBH.LA1	Load bearing – dry conditions
MBH.LA2	Heavy duty load bearing – dry conditions
MBH.HLS	Load bearing – humid conditions
MBH.HLS2	Heavy duty load bearing – humid conditions
Softboards	
SB	General purpose – dry conditions
SB.H	General purpose – humid conditions
SB.E	General purpose – exterior conditions
SB.LS	Load bearing – dry conditions
SB.HLS	Load bearing – humid conditions
Medium density fibreboard (MDF)	
MDF	General purpose – dry conditions
MDF.H-1 and MDF.H-2	General purpose – humid conditions
MDF.LA	Load bearing – dry conditions
MDF.HLS-1 and MDF.HLS-2	Load bearing – humid conditions

an imported Scandinavian Timber Frame House kit. And as I said, softboard is still being referred to as 'Insulation Board', despite it no longer having any superior insulation qualities when compared to many more modern materials on the market.

10.21   MDF

These initials actually stand for 'Medium Density Fibreboard' – but nobody ever calls it that, which is just as well, since that would then risk confusion with the 'medium density' grade of Wet Process fibreboard! (I know what I've just said might sound a bit daft on first reading: but the two panel types are very different products, despite their having confusingly similar names.)

MDF is in fact made by the Dry Process: and for this reason it looks and behaves very differently from all of the other fibreboards that I've just mentioned. Let me explain.

The Wet Process boards all have one shiny face and one 'coarse' face – which has a 'mesh' pattern impressed into it. That's because, as part of the manufacturing process, the wet fibre mat that these boards are made from is allowed to drain off quite a lot of its excess water – a bit like straining tea, in fact – by forming the wet mat on a wire mesh belt. In this way, the boards all end up with an impression of this 'strainer' mesh in their back faces. Therefore, the Wet Process boards are more restricted in what they can be used for, because the mesh face is more absorbent and it doesn't look particularly attractive.

But the Dry Process, on the other hand, uses completely dry fibres (surprise, surprise!) that are glued together in the same manner as are the particles in wood chipboard. So MDF does not need to be formed on a mesh belt: it can have a shiny steel belt instead. Therefore MDF has two shiny faces, rather than one shiny face and one mesh face. But it also has another highly useful attribute that chipboard doesn't possess: machineability.

You will all, I'm sure, have seen veneered furniture with spray-lacquered, moulded (machined) edges; but perhaps you were not aware that MDF was the substrate being used to carry the decorative surface veneer. The reason why MDF has been so successful and so popular in furniture-making and shopfitting in recent years, is precisely that ability to be machined into complicated and curved profiles, in the way that solid wood can be – and in a way that chipboard can't be (see Figure 10.7).

So with MDF you get a very large, flat and highly stable panel that can in many ways be used just like timber. Possibly its only drawback is that it is very dense and therefore very heavy. But nowadays there are lightweight and even 'bendy' versions available for joiners and shopfitters to play with.

The only thing that MDF cannot do well is to be used fully out of doors. Look at Table 10.3 and you'll see: there is no 'Exterior' version of MDF under EN 622-5. (Yes, I know that there *is* an MDF product that calls itself 'Exterior': but if you read the small print in the manufacturer's brochure, you will see that it only fits the 'Humid use' category and you will see that it must be 'protected' when used externally. So it's not really fully Exterior, is it?)

Well, that's the three different families of wood-based boards covered in their essentials. Now I'll recap the vital bits.

Figure 10.7   Moulded MDF – can be machined as solid wood

10.22   Chapter summary

I have described all the vital stuff that you need to know about the three different 'families' of wood-based sheet materials: plywood, chipboard (including OSB) and fibreboard (including MDF); and how they are made, using different and ever-more reduced forms of wood. From veneers in plywood to particles in chipboard (or strands in OSB) and wood fibres themselves, in the last of these types of wood-based boards.

Regarding plywood, I described how it is more dimensionally stable and more uniform in its strength than solid wood. And I listed the different basic types: Conifer, Temperate Hardwood and Tropical Hardwood. I advised you to use only Certified and Quality-Assured plywoods for any serious building tasks; and I tried to warn you off using un-certified and essentially unreliable Tropical Hardwood plywoods.

And I definitely told you that 'WBP' plywood does not exist – since it has been declared obsolete for years – and so for any uses of plywood where delamination could be a real problem, you should ask for it to be 'Class 3 Exterior' to BS EN 314-2.

Regarding particleboards, I told you that chipboard is the wood-based version of them: and there are only a few types which the building industry is really interested in: mainly the moisture resistant flooring type known as P5. Chipboard is much more reliable and trustworthy in its manufacture and therefore its quality than is plywood; since most of what we use is made either here in the UK, or in Europe. And remember that there is no fully Exterior form of particleboard.

OSB is also a sort of particleboard: but its properties make it much more akin to 3-ply plywood: and because of that it has taken over many of the building uses that plywood formerly did.

And so to the fibreboards. Hardboard is really the only one of the more 'traditional' type that you will see. But of course, MDF (which is a 'Dry Process' board) is now found virtually everywhere: and this is thanks to its ability to be machined and moulded like solid wood.

Tempered Hardboard is the only Exterior type of fibreboard: and there is no fully Exterior quality of MDF listed in the European Standard.

So that's it. Now have a look in the Glossary for some useful explanations of terms used with timber in construction; and look in the Appendices for more help and some other useful references.

Above all – use wood properly and with a better understanding, and it will last you well and look good for the life of your building.

Appendix 1

A Glossary of Wood and Timber Terms Used in the Timber and Construction Industries

This Glossary is not intended to be an exhaustive 'Timber Dictionary' – but it is meant to help unscramble some of the rather arcane terms which the Timber Trade likes to use to confuse outsiders!

ACQ:	A type of wood preservative, based on Alkalised Copper Quaterany ammonium Compounds. It is impregnated under pressure and it is one of the 'new generation' wood preservatives that have replaced CCA in many areas of use.
Air dried:	The state reached by timber when in equilibrium with natural atmospheric conditions out of doors (also called 'seasoned'): in the UK, about 15% mc in summer and 18% mc in winter.
Annual ring:	See Growth Ring.
Arris:	The corner of the length of a sawn or planed section of timber, where the face and edge meet.
Arris rail:	An originally square section of timber, sawn diagonally across the section to make two triangular-section pieces. Used in fencing.
Band saw:	A continuous steel belt running between large diameter pulley wheels, having saw teeth on the leading edge. Large-bladed (i.e. 150 mm and up) bandsaws are used for converting logs, and smaller saws are used for resawing timber into smaller sections.
Barge boards:	Sloping pieces of timber which trim the gable ends of a pitched-roof building.

Wood in Construction: How to Avoid Costly Mistakes, First Edition. Jim Coulson.
© 2012 John Wiley & Sons, Ltd. Published 2012 by John Wiley & Sons, Ltd.

Batten:	A piece of square-edged softwood; ranging from 50 mm to 100 mm thick and 100 mm to 200 mm wide.
Beam:	A large and heavy section of timber used horizontally; often 150 mm by 250 mm or greater: used to carry the joists in a floor construction. Sometimes with an adzed, rather than a sawn, finish.
Bill of Lading:	A known specification of timber, for which a Shipper produces documentation itemising the number of pieces, sizes and lengths, usually as ordered by a Timber Importer.
Blue stain:	Bluish or blackish discolouration which penetrates the surface of susceptible woods and causes loss of visual quality. It is caused by certain fungi that are *not* rots: so blue stain does not break down the wood and is therefore not a structural defect.
Board:	A piece of square-edged softwood; under 50 mm thick and over 100 m wide.
Bow:	Distortion of timber by curving flatwise along its length.
Bracking:	The term used for sorting timber into grades: usually when referring to softwood of Scandinavian or Russian origin.
Carcassing:	The term for rough-sawn or hit-and-miss planed timber used to construct the shell or carcass of a building: i.e. the roof timbers, joists and wall studs, together with any extra non-loadbearing pieces, such as noggings or firrings.
CCA:	Copper-Chromium-Arsenic wood preservative: based on inorganic salts of these chemicals, dissolved in water and impregnated into timber under pressure.
Check:	A drying defect in timber, taking the form of a very slight separation of the fibres along the grain.
	Alternatively, a defect of similar appearance in plywood; but caused by the action of peeling and 'unrolling' a veneer from the log: often called 'lathe checks'.
Chipboard:	A wood-based sheet material (particleboard) made from particles or chippings of wood, glued together and pressed into a flat panel.

Clear:	Descriptive of timber which is free from large knots and other visual defects, such as resin or bark pockets.
Clears:	A top visual Quality from North America and found in the PLIB Export 'R' List.
Conversion:	The sawing or otherwise of a log into (usually) rectangular cross-sectional pieces of timber.
Core Veneer:	One of the centre veneers in plywood: often thicker than the face veneers, especially in Tropical Hardwood plywood.
Country Cut:	(*no longer a common term*) Timber supplied as sawn in the imported size and not having been resawn from a larger original dimension.
Cross-Cut:	Timber sawn at right angles to the grain across the end of the piece or the log.
Crosscut Saw:	A saw specifically for cross-cutting timber: usually circular.
Crown Cut:	A method of cutting veneers tangentially, so that the production of 'flame' figure is maximised.
Cupping:	Distortion of timber by curving flatwise across its width.
Cut roof:	A traditionally-made, hand-built roof, using a ridge board, purlins and common rafters (but not trussed rafters or roof trusses).
DAR:	Dressed all round: a Scottish term, the same as PAR.
Deal:	A piece of square-edged softwood; ranging from 50 mm to 100 mm thick and 225 mm to 275 mm wide. Alternatively, an old-fashioned term for any softwood (eg: Redwood is sometimes called 'red deal').
Decay:	The destruction of the structure of wood by rotting fungi, which can result in loss of structural strength. Commonly called 'Rot' (sometimes, as with white pocket rot, also called 'Dote').
Defect:	A fault or irregularity (due either to growth, conversion or drying) which detracts from the strength or appearance of a piece of timber or wood-based panel.

Dote:	See Decay.
Dressed:	Same as Planed.
Eased edge/arris:	The slight rounding of an edge or arris of a square-sawn section of timber, done to improve handling or finishing properties.
emc:	Equilibrium moisture content (usually written in lower case): the mc achieved by any wood product when in balance with the prevailing atmospheric conditions – indoors; unheated, etc.
Elliottis Pine:	The species *Pinus elliottii* grown in plantation in the Southern hemisphere, often used in plywood production. (Not to be confused with Southern Pine.)
Ex larger:	The usual term to denote that a piece of wood has been resawn from an original (imported) size into two or more smaller sections. There will always be a slight size difference between the full original size and the resawn size, owing to the sawcut width (i.e. the kerf).
Face Veneer:	The front or back surface of a sheet of plywood: sometimes having a decorative figure. (In many cases referred to by letters such as B/BB; II/III, etc.)
FAS:	Firsts and Seconds: the top appearance quality grouping of many imported hardwoods, especially those from the USA.
Fascia:	A horizontal board, usually made of timber or plywood, which trims the eaves of a roof: often carrying the guttering.
Feather edge(d):	Timber cut so that it tapers in its cross-section; with one edge thinner than the other. F/e boards are often seen as horizontal exterior cladding, with the thicker edge at the bottom.
Fibreboard:	A family of wood-based sheet materials made from compressed wood fibres, usually without adhesive (*Wet Process*); except for MDF (*Dry Process*).
Fifths:	The visual quality of Scandinavian timber below Unsorted, generally deemed suitable for carcassing or lower quality joinery.

Alternatively, the visual quality of Russian timber below that country's Fourths. Used for the same purposes as Scandinavian sixths.

Figure: The decorative appearance on the surface of timber: it may be caused by growth ring pattern, grain irregularities, or other factors in the wood's make-up, or its conversion.

Finished size: The actual size of a piece of timber after any sawing, planing or moulding.

Fillet: A piece of wood which tapers along its length. Fillets (also known as 'firring pieces') are fixed on top of flat roof joists in order to create a slight fall for the roof covering.

Firring Piece: Same as a fillet, above.

Flange: One of the two outer edge members of an I Beam or Box Beam.

Fourths: Formerly, the lowest visual quality of Scandinavian timber deemed suitable for joinery work. Now incorporated as part of the Unsorted bracking.

Alternatively, the visual quality of Russian timber below that country's Unsorted quality. Generally used for the same purposes as Scandinavian fifths.

FSP: Fibre Saturation Point: the moisture content at which all the liquid water has dried out of a piece of timber, and only the cell walls are still fully wet. From this level downwards, timber is subject to shrinkage, but above this level, it will not swell up. Reckoned to be about 28% mc for most wood species.

Gauging: Same as thicknessing or regularising.

Grading: A method of selecting different qualities or grades of timber, usually based on allowable amounts of visual defects within the range of pieces coming off the saw. May be done by trained operatives, or by a special machine.

Grain: The direction of the elements of growth in the tree or board.

Growth ring: The ring seen on the end-grain of a log or piece of timber: in temperate woods, made up of earlywood and latewood tissue (these being the two parts of cell growth during a season). In temperate timbers, the growth rings are

annual rings and will correspond with the age of the tree: but in tropical timbers, there is no correlation between the number of rings and the tree's age.

Hardboard: The highest density type of fibreboard. Often further finished by processing with boiled linseed oil, to produce **tempered hardboard**.

Hardwood: Any timber obtained from a broad-leaved tree. Purely a botanical term; so that a 'hardwood' is not necessarily 'hard' to the touch (e.g. Balsa, which is botanically a 'hardwood').

Humidity: See Relative Humidity.

Joist: A sawn (and commonly, processed or 'regularised') timber member which supports the floor and ceiling in a building.

KAR: Knot Area Ratio. Used in the assessment of Strength Graded softwoods to BS 4978.

Kerf: The width of wood removed by the sawcut.

Kiln dried: The state of timber which has been subjected to artificial drying, as opposed to air seasoning. Imported softwoods are usually partially kiln dried to remove excess moisture prior to shipping. For timber in service in heated buildings, kiln drying is essential to achieve a low enough moisture content to avoid problems of splitting, shrinkage and distortion.

Knot: A defect in timber caused by the cut-through portion of any branch showing on the sawn or planed surface. Knots will affect the surface appearance of wood in *quality* grades; but the size and distribution of knots is also a very important consideration when assessing the load-bearing capacity of timber in *strength* grading.

Lath: A very thin strip of wood (usually about 6 mm thick by 25 mm wide) fixed to a wall or ceiling as a base for plaster.

                    Alternatively, the same as a 'sticker' used in air drying.

Length packaged: See packaged to length.

Log:	The round trunk of the tree, without branches, which is converted into timber at the sawmill. Usually, it is only a section of the entire tree trunk, varying in length from about 4.5 to 7.0 m.
	Alternatively, a large dimension piece of square-edged timber, often up to 350 mm by 350 mm section and up to 12.0 m long: used for beams or for resawing into larger joists and longer lengths. In softwoods, generally only available from Western hemlock and Douglas Fir.
Lumber:	The North American term for 'Timber' (i.e. sawn wood).
Matchboarding:	Tongued and grooved boarding used as interior cladding or panelling; often with v-jointed edges.
mc:	Moisture Content (usually written in lower case). Expressed as a percentage of the *Dry Weight* of any given piece of timber.
MDF:	Medium Density Fibreboard (usually just called MDF, to avoid confusion with mediumboard). A wood-based sheet material pressed from wood fibres with a small amount of added glue: it has good flatness properties and it can be moulded and shaped like solid wood.
Mediumboard:	A medium-density quality of fibreboard, akin to hardboard although not so dense. Not to be confused with MDF.
Merchantable:	Qualities of Canadian timber (usually referred to as 'Merch') which contain some defects, but which are suitable for more general work. The usually available qualities from the Export 'R' List are: Select Merch, No. 1 Merch and No. 2 Merch.
M/F Resin:	Melamine-Formaldehyde based plastic resin adhesive; used in the manufacture of moisture-resistant grades of chipboard, etc.
Moisture content:	See mc.
Moulding:	The shaping of timber by means of a planing machine, using specially-shaped ('profiled') cutters.

Alternatively, any piece of shaped timber as produced by the moulding process.

Muntin: The central vertical member of a panelled door, which separates the panels, running between the top rail and lock rail, and/or lock rail and bottom rail.

Noggin(g): A small-section piece of timber (50 mm × 63 mm or 50 mm × 75 mm) nailed between studs, to provide a base for fixing items to.

Nominal Size: The quoted size of timber as originally sawn or produced, which is not necessarily the actual finished size of the component. This may be due to subsequent shrinkage, or other processing, such as 'regularising'.

Oil-Tempered Hardboard: See hardboard.

O/S: Organic Solvent-based wood preservatives: complex organic chemicals, often dissolved in a white spirit-type carrier.

OSB: Oriented Strand Board. A composite wood-based panel product, made from strands or flakes of timber, with the strands in alternate layers aligned approximately at right angles to one another. It may be used structurally.

PAR: Planed all round: denotes timber which has had all four surfaces dressed.

P/F Resin: Phenol-Formaldehyde plastic resin adhesive; used in the manufacture of Exterior plywood, OSB, etc.

Packaged to length: A package of timber, usually 1200 mm wide and 1200 mm high, with all pieces in the pack cut to the same standard length.

Peeling: A method of veneer production: see Rotary Cut.

Plywood: A manufactured sheet material consisting of veneers of wood glued to one another, usually with alternate layers (plies) having the grain running in alternate right-angled directions.

Post:

A vertical structural timber member, larger than a stud, often square in section. Used in conjunction with beams.

Alternatively, a timber (75 mm × 75 mm, or 100 mm × 100 mm square: or 75 mm to 100 mm round section), pointed at the bottom end, driven into the ground, and used to support the rails in fencing.

PSE:

Planed, square-edged: often taken to mean the same as PAR; but may have two planed faces, with square-sawn edges.

Purlin:

A large section piece of timber (generally at least 75 mm × 250 mm) which supports the common rafters in a cut roof.

Rafter:

A sawn timber member which supports the roof covering; usually fixed at a sloping angle of between 15° and 45°.

Rail:

A horizontal member in a timber frame construction, running across the tops of the posts or studs.

Alternatively, one of the horizontal members in a panelled door: usually divided into Top rail, Lock (middle) rail and Bottom rail.

Alternatively, a horizontal member used in fencing, nailed to the posts, to provide a barrier (for livestock, etc).

Regularising:

The action of sawing or planing two opposing edges of a piece of timber, to ensure uniform thickness: often applied to floor joists to give a level floor surface. Same as gauging or thicknessing: but nowadays superseded by the term 'processing'.

Relative Humidity:

Abbreviated to 'RH'. The amount of water vapour which air at a given temperature can hold. Warm air holds more water than cold air: so that as air temperature drops, condensation occurs, and as air

	temperature rises, evaporation takes place. The RH of the in-service atmosphere is the main factor in deciding the end-use moisture content of any timber item.
Resawing:	The act of sawing timber in its length, to produce two or more smaller sections from an originally larger size.
RH:	See Relative Humidity.
Ridge board:	The board at the apex of a cut roof, where the top ends of the rafters meet together at an angle.
Ring Shake:	A separation of the fibres around the curve of the growth ring: caused by damage to the living tree.
Roof truss:	A manufactured construction having bolted or carpenter-made joints and with the members *overlapping*. It is used to support smaller purlins and common rafters in a larger, older type of engineer-designed roof. (Do not confuse this with a *trussed rafter*.)
Rot:	See decay.
Rotary Cut:	A method of cutting veneers so that the log is 'unrolled' along the axis of the growth ring. More commonly called 'Peeling'.
Sarking:	Boards fixed over the rafters of a cut roof or trussed rafter roof, to provide stability. May be softwood, but these days more usually plywood or OSB.
Saw Falling:	The quality of Scandinavian softwood sold without any pre-selection or grading; just as it 'falls off the saw'. Sometimes sold as a quality which excludes Sixths.
Seasoned:	An older term for Air Dried. This term can give the impression that the timber is now 'stable', but this is not so: it can still react to changes in environmental conditions – outdoors to indoors, etc.
Select:	A very high quality appearance grade of Canadian timber.
Set:	The way in which saw teeth are positioned, in order that the saw does not 'bind' in the sawcut. Circular crosscut and hand-saws are 'spring set' and bandsaws are usually 'swage set'.

Sevenths:

The lowest quality of Scandinavian softwood; where any defect is allowed. Not usually found as an Export quality.

Shake:

A larger form of check (a separation of the fibres along the grain) often penetrating into the centre of a square-edged piece of timber: usually resulting from drying too quickly.

Shipper:

The producer of timber – often, but not always, the saw-mill – in the country of production or export.

Shipper's usual:

The normal run of quality expected for any given appearance bracking, when bought from *one particular Shipper*. (NOTE: the same visual quality grade, when bought from two separate Shippers, either from different countries, regions or even mills, may not be the same in actual quality or appearance; despite both Shippers working to the same basic grade description.)

Sixths:

A visual quality of Scandinavian timber below fifths; generally deemed suitable for lower quality uses, such as packaging and fencing. In Sweden, called 'UTSKOTT'.

Slope of Grain:

The angle by which the grain direction deviates from 'straight' (i.e. parallel with the length of the piece). Excessive slope is a structural weakness in strength graded timber, and any slope of grain may cause distortion in the timber on drying.

Soffit(e):

The underside of any part of a constructional element: especially of a ceiling or eaves.

Softwood:

Any timber obtained from a coniferous tree. Purely a botanical term: so that a 'Softwood' is not necessarily 'soft' to the touch.

Spiral Grain:

A form of Slope of Grain, caused by the fibres in the living tree growing at an inclination to the vertical, so that they grow both up and around the trunk in a shallow spiral. This can give rise to *twist* when sawn boards are dried.

Split:

A defect in timber caused by a separation of the wood fibres along the grain, generally penetrating fully through the piece.

Spring:	Distortion of timber by curving sideways along its length.
Spring Set:	See 'Set'.
Stick(er):	A thin piece of timber (usually 12 to 20 mm thick) inserted between the layers of wood in a pack, in order to assist drying and/or to help stabilise packs of timber by binding them together. (Also may be called a 'lath' a 'binder' or a 'pin' in various parts of the UK.)
Strength Grading:	A method of assessing timber for structural strength, based on its inherent defects; but taking much less account of its actual visual quality. Strength grading may be done visually (by trained and certificated personnel) or by specially programmed machines. All strength graded timber must be stamped with a quality assurance mark or code indicating its grade, together with other identifying information, such as mill, grader and species.
Stress Grading:	The former term for Strength Grading (see above).
Strut:	A small section of timber braced across between joists to prevent sideways rotation. Sometimes paired in an 'X' formation: then known as 'Herringbone Strutting'.

Alternatively, an internal member in a truss or trussed rafter, designed to take compression forces. |
Stud:	A vertical member which forms part of the structural framing of a timber framed wall: it may or may not be load-bearing.
Swage Set:	See 'Set'.
T&G:	Tongue and Groove: a joint profile used in flooring, panelling, cladding, etc.
Tanalith/Tanalising:	A proprietary brand of CCA – or other similar chemical formulation – wood preservative; and its process.

Tempered Hardboard:	See hardboard.
Tie:	An internal member in a truss or trussed rafter, designed to take tension forces.
	Alternatively, any member acting principally in tension: such as a tie beam around the perimeter of a floor.
Thicknessing:	The same as regularising. Now replaced by the term 'processing'.
	Alternatively, another term for planing.
Truck bundled:	A package of timber, usually 1200 mm wide by 1200 mm high, containing pieces of random length, with only one end of the pack squared off.
Trussed rafter:	A factory-made roof component in which all pieces are jointed *in the same plane* using punched metal plates. It must be constructed into a roof using either bracing and/or sarking. (These items are often referred to simply as a 'trusses', but they should not be confused with proper *Roof Trusses*).
Twist:	Distortion of timber in a shallow spiral along its length.
U/F Resin:	Urea-Formaldehyde plastic resin adhesive; used in the manufacture of chipboard, MDF, etc.
Unsorted:	A visual quality of Scandinavian timber consisting of the top four grades, not otherwise sorted or sold separately. Normally used for high class joinery, cladding, etc, where appearance is most important.
	Alternatively, a visual quality of Russian timber consisting of that country's top three grades, not otherwise sorted or sold separately.
V-jointing:	The machining of a bevelled edge to the tongue and groove butt joints of matchboarding which is done to emphasise the joints.

Veneer: A thin slice of wood, forming one of the 'plies' in plywood. It may range from less than 1 mm thick, up to 3 or 4 mm in thickness, depending on its type and position: ie, core or face. Veneers are also used as decorative overlays onto solid timber or onto other wood-based board substrates.

Wane: A defect in sawn timber, caused by producing rectangular timber from tapering, round logs. One or more arrises may then be partially or wholly rounded, instead of being square-edged.

WBP: 'Weather and Boil-Proof' – an obsolete term used in relation to Exterior grade plywood (formerly in the now-withdrawn BS 6566). It has long since been replaced by 'Class 3 Exterior' to BS EN 314-2.

Web: The internal member(s) of an I Beam or Box Beam.

XLG: 'X-Ray Lumber Gauge' – a type of strength grading machine (developed in Canada) which uses X-Rays to asses the density of timber being graded. It can accurately identify knots and areas of low strength; but it cannot 'see' slope of grain.

Yellow Pine: Either Quebec Yellow Pine (*Pinus strobus*) or Southern Pine (principally *Pinus elliottii*). The latter is often – misleadingly – called 'Southern Yellow Pine', but the two species are quite different in their properties and appearance and should not be confused.

Appendix 2

A Select Bibliography of Some Useful Technical References About Wood

Building Research Establishment (1975) *A Handbook of Hardwoods*, HMSO.

Building Research Establishment (2nd edn, 1979) *A Handbook of Softwoods*, HMSO.

Building Research Establishment (3rd edn, 1997) *Timber Drying Manual*, HMSO.

Desch H.E. (Revised Dunwoodie J.) (6th edn, 1983), *Timber Its Structure and Properties*, Macmillan.

Hoadley R.B. (2nd edn, 2000), *Understanding Wood: A Craftsman's Guide to Wood Technology* (2nd edn), Taunton Press.

Oxley T.A. and Gobert E.G., (2nd edn, 1998) *Dampness in Buildings*, Butterworths.

Wood in Construction: How to Avoid Costly Mistakes, First Edition. Jim Coulson.
© 2012 John Wiley & Sons, Ltd. Published 2012 by John Wiley & Sons, Ltd.

Appendix 3

Some Helpful Technical, Advisory and Trade Bodies Concerned with Timber

I'm sure you're computer-literate enough to be able to put these names into a Search Engine on your own computer and find out full contact details for them.

BM TRADA Certification	A leading Certification Body concerned with Chain of Custody, Sustainable Timber Supply and Auditing of such systems: also the licensing of timber graders.
Technology For Timber Ltd TFT Woodexperts	A consultancy practice helping to solve problems with wood in service; assisting with timber design; and undertaking site investigations. Also delivering training on timber to the Timber Trade and the users of wood and wood-based materials.
AHEC	The American Hardwood Export Council: a very helpful body providing high-quality advice on USA Hardwoods and their grades.
Timber Trade Federation (TTF)	The Trade Body representing many of the leading Importers and Merchants of Wood Products within the UK.
British Woodworking Federation (BWF)	The Trade Body representing many of the leading manufacturers of joinery products within the UK.
Wood Panel Industries Federation (WPIF)	The Trade Body representing Manufacturers of Wood-based Panels within the UK and Ireland.

Wood in Construction: How to Avoid Costly Mistakes, First Edition. Jim Coulson.
© 2012 John Wiley & Sons, Ltd. Published 2012 by John Wiley & Sons, Ltd.

Index

Abies alba 141
Acer saccharum 157
ACQ preservatives 128, 186
across-the-grain 6–8, 17, 48–9
adhesives 171–2
African Mahogany 156
air drying 43, 53–4, 56, 62–3, 186
alkalised/ammoniacal copper quaternaries
    (ACQ) 128, 186
along-the-grain 6–8, 17
American Whitewood 163
anaerobic conditions 73
Angiosperms 20
APA quality stamps 176–7
appearance grading 83–7, 99, 173–7
Araucaria angustifolia 21, 28–9, 147–8
architectural grade 116
Ash 151–2

Bagassa guianensis 161
bagasse board 178
Baltic Pine 106
Beech 152–3
Betula pubescens 153
Birch 153
Birch plywood 167, 168, 175
blue staining 61, 105, 106, 187
borer hole no defect (BHND) 99
boron-based preservatives 124–5
British Spruce 111–12, 115, 141–2
British standards 27–30, 37
    moisture content 44
    strength grading 107–11, 115–16
    use specification 57–8, 67–8, 74, 78
    wood-based sheet materials 167, 170–2,
        174, 178, 181–3
broadleaf trees 20–1
    see also hardwoods

build of finishes 135
Building Regulations (UK) 109–10
Building Research Establishment (BRE)
    117–18

cabinet making 97, 147, 153, 157
CanPly quality stamps 176
Castanea sativa 153–4
CCA preservatives 126–7, 137–8, 187
Cedar 47, 68–9
cell structure 19, 23–7
cellulose 2–4, 17
Central American Mahogany 69–70
central heating 53–4
Certificates of Grading 119–20, 121
certification of plywoods 176–7
Chain of Custody 70
Cherry 153
Chestnut 117, 153–4, 159
chipboards 2, 177–81, 187
clear cutting system 83, 93–103
clear timber 87, 92–3, 95–6, 188
colour 11–13
comb grain 144
commercial forestry 18
commons grades 92–3, 96
compression wood 84–5
comsel grade 97–8
condensation 64
conifer plywoods 166–7, 172–5
conifers 19–20
    see also softwoods
Coniophora puteana 75
construction industry 19
    hardwoods 155–6, 159
    moisture content 48–9, 53–4
    quality and grading 80–1
    softwoods 140, 142–3, 147, 149

Wood in Construction: How to Avoid Costly Mistakes, First Edition. Jim Coulson.
© 2012 John Wiley & Sons, Ltd. Published 2012 by John Wiley & Sons, Ltd.

strength grading 104–7, 109–16, 120–1
use specification 62–6, 69–70, 75–7
wood-based sheet materials 179
copper, chromium and arsenic (CCA)
    preservatives 126–7, 137–8, 187
cordons sanitaire 122
cracking 62–3, 80–1
cross-banding 165–6
cross-sectional area 117–18
cutting-based grading 83, 93–103

decay 52, w188
deck boards 67–8, 146, 161
defect-based grading 83–93, 102, 188
delamination of veneers 169
delivery care 2
depth of penetration 69, 73, 122–3
desired moisture content 44–6
desired service life 69, 73–5, 77–8
diffuse-porous hardwoods 25–6, 36–7
dimensional changes 8–10, 17
Dipterocarpus spp. 156
direction of loading 4, 6–8, 16
distortions 84–6
double vacuum method 125
Douglas Fir 77, 91, 143–4
Douglas Fir-Larch (DFL) 149
dry wood 41–4, 50–1
drying timber 5
durability
    finishes 135
    hardwoods 154–5, 158–63
    preservative treatments 124, 131
    quality and grading 79–80
    softwoods 141–2, 145–6
    use specification 58–9, 65, 67–70, 73, 77

earlywood 24, 32–3, 35, 144
Ekki 154
electrical resistance-type moisture
    meters 40–3
Entandrophragma spp. 160–2
environmentally-friendly preservatives 127–8
equilibrium moisture content (EMC) 44–6,
    51–6, 62, 65, 189
European redwood 74, 77, 80, 115, 139–40
European standards 27–30, 37
    appearance grading 90
    moisture content 44

quality and grading 100–3
strength grading 108–9, 112–13, 119–20
use specification 57–8, 67–8, 78
wood-based sheet materials 167, 170–5,
    178–84
European Walnut 162
European whitewood 115, 141–2
expected moisture content 44–5
Export R List 90–2, 94
exposed face 101–2
external cladding 67–8, 137, 145–7
extractives 12–14, 17

F1F grade 97
face quality 172–3, 189
Fagus sylvatica 152–3
fence posts 70–4
fence rails 67
fibre saturation point (FSP) 46–7, 190
fibreboards 2, 181–5, 189
fibres 24
fifths 88–9, 93, 189
figure (grain pattern) 4–5, 157, 189
film-forming paints 131–3
finishes 129–38
    durability 135
    effects of light/dark colours 136–7
    material properties 2
    microporous finishes 133, 137
    outdoor use 130–5
    paint 2, 131–5
    stains 2, 133–4, 137
    varnish 131–3
first and seconds (FAS) 96–7, 189
fitness for purpose 82
flaking 132
flaxboard 178
floor joists 75, 106, 109–10, 115, 191
forestry 18, 70, 87, 106, 148
fourths 89, 190
Fraxinus spp. 151–2
fungal damage 11–13, 64, 67, 70, 75
furniture making 152–3, 155–7, 159–62

general structural (GS) grade 107–8, 112–13
glue bond 169–70
Grading Certificates 119–20, 121
grain 4–10, 16, 165–90
green wood 47–8

Greenheart 77, 154
Gribble 76–7
growth rings 10–11, 31–2, 33–8, 190–1
Gymnosperms 20

hammer probes 42–3
hardboard 181–3, 191
hardwoods
    cell structure 24–6
    definition and characteristics 19–23, 191
    diffuse-porous 25–6, 36–7
    geographical distribution 21–3, 31–2
    growth rings 31–2, 37–8
    material properties 1–2, 12
    moisture content 48, 54–5
    movement 48
    quality and grading 80, 83, 94–103
    rate of growth 34–8
    ring-porous 25, 34–6
    strength grading 115–19, 120–1
    timber trade 19–27, 31–2, 34–7
    United Kingdom species usage 150–63
    use specification 58, 65, 77
    wood-based sheet materials 153, 157,
        168–9, 172–3, 175–7
hazard 60–1, 65, 78
heartwood 10–13, 17, 144, 146–7
Hem-Fir 142, 149
high density mediumboard 182
high pressure treatment 126, 129

Idigbo 155
improved face veneer 174
in-service moisture content 43–4, 46, 55,
    59–60, 64–5
indoor use 61–6, 147, 152–3, 163
insect damage 11–13
insecticides 128
Iroko 12, 65, 70, 155

joinery 143–4, 147–8, 155–7, 163
joinery (J) classes 100–1, 103
Juglans spp. 162

Keruing 156
Khaya spp. 156
kiln drying 39, 43, 49–53, 56,
    62, 191
knot clusters 80–1, 84

Larch 145
Larix spp. 145
latewood 24, 32–3, 35, 144
lengthwise movement 8–9
lignin 6–7
Limnoria spp. 76–7
Liriodendron tulipifera 163
lock gates 70–1
log ends 10–13, 15
Lophira alata 154
low density mediumboard 182
low pressure treatment 125

machine grading 110–12
machining 9–10, 30
Madison Formula 133
Mahogany 12, 69–70, 156–7
Malaysian Grading Rules (MGR) 94,
    98–9, 103
Maple 157
marine construction 76–7
marine plywood 171–2
marking systems 119–20, 121
material properties 1–17
    cellulose 2–4, 17
    colour 11–13
    dimensional changes and movement
        8–10, 17
    direction of loading 4, 6–8, 16
    grain and figure 4–10, 16
    heartwood and sapwood 10–13, 17
    machining 9–10
    moisture content 2, 8–10, 11
    natural durability 13–14, 17
    permeability 15–16, 17
    radial and tangential directions 15
    rays 14–16, 17
    temperature changes 8
    timber quality and grade 2
    wood species differences 1–2, 11–12
medium boards 181–3, 192
medium density fibreboard (MDF) 2,
    183–5, 192
melamine formaldehyde (MF) 171, 192
Meranti 60, 157
merchantable grades 92–3, 192
microporous finishes 133, 137
Milicia excelsa 65, 70, 155
moisture content 39–56

air drying 43, 53–4, 56
  central heating 53–4
  decay safety limit 52
  definition and calculation 39–40, 192
  desired moisture content 44–6
  equilibrium moisture content 44–6,
    51–6, 62, 65
  fibre saturation point 46–7
  finishes 137
  hardwoods 153
  heartwood and sapwood 11
  in-service moisture content 43–4, 46, 55,
    59–60, 64–5
  kiln drying 39, 43, 49–53, 56
  material properties 2, 8–10, 11
  moisture meters 40–3, 55
  movement 46, 48–9
  shrinkage 46, 47–8
  specification 44–6
  timescales for drying timber 54–5
  use specification 59–60, 62–77
moisture meters 40–3, 55
movement
  material properties 8–10, 17
  moisture content 46, 48–9
  softwoods 142
  timber trade 30–1
  use specification 62

National Hardwood Lumber Association
    (NHLA) 94–8, 103
National House Building Council
    (NHBC) 65–6, 74
natural defects 84, 99
Nauclea diderrichii 65, 77, 160
non-film-forming paints 133–4
North American grading 90–3, 102, 112–13
Norway Spruce 141–2

Oak 12–13, 55, 65, 116–18, 158–9
Obeche 159–60
Ocotea rodiaei 77, 154
Opepe 65, 77, 160
organic preservatives 128, 137–8, 193
oriented strand board (OSB) 178–81, 193
outdoor use 66–77, 130–5, 145–6,
    161, 170–1
oven dry 39–40
over-drying 62–3

Pacific Lumber Inspection Bureau (PLIB)
    90–2, 94
paint 2, 131–5
Parana Pine 21, 28–9, 147–8
parenchyma 26, 37
particleboards 177–81, 185
peeling 132, 193
permeability 15–16, 17, 122, 142, 147
permissible defects 84, 99
phenolic resin (PF) 171, 193
photosynthesis 3
Picea spp. 122–3, 141–2
pin hole no defect (PHND) 99
Pine
  material properties 13
  preservative treatments 123, 125
  strength grading 106, 115
  timber trade 21, 28–30
  United Kingdom species usage 140,
    142, 146–7
  use specification 74
Pinus elliottii 30–1, 146–7, 189
Pinus palustris 146–7
Pinus strobus 28–31, 147
Pinus sylvestris 74, 77, 139–40
pinworm 81, 106
Pitch Pine 146
pith streaks 80
pits 16
planed all round (PAR) 193
planed timber 4
plywood 153, 157, 164–77, 185, 193–4
  adhesives 171–2
  appearance grading 173–7
  basic types 166
  conifer plywoods 166–7, 172–5
  cross-banding 165–6
  face quality 172–3
  fundamental properties 165–6
  glue bond and WBP 169–70
  hardwoods 168–9, 172–3, 175–7
  marine plywood 171–2
  material properties 2
  outdoor use 170
pores 24–6, 34–7
preservative treatments 122–9, 137–8
  basic methods of timber treatment 124–6
  CCA preservatives 126–7, 137
  cordons sanitaire 122–3

preservative treatments (*cont'd*)
  depth of penetration 122–3
  environmentally-friendly
    preservatives 127–8
  high pressure treatment 126, 129
  low pressure treatment 125
  material properties 2, 13–14, 16, 17
  organic preservatives 128, 137–8
  permeability 16, 17
  strength grading 115
  Tanalised timber 128
  treated timber specification 129
  types of preservative 123–4
  use classes 123–4, 127–8, 137
  use specification 58–9, 65–6, 68–70, 73–5,
    77–8
prime grade 97, 98–9
Prunus serotina 153
Pseudotsuga menziesii 77, 91, 143–4

quality and grading 79–103
  appearance grading 83–7, 99
  BS EN 942 standard 100–3
  clears, merchantable and commons 92–3
  cutting-based grading 83, 93–103
  defect-based grading 83–93, 102
  distinction between quality and
    grade 81–3
  European appearance grading 90
  exposed face 101–2
  F1F grade 97
  FAS, selects and commons 96–7
  fitness for purpose 82
  grade defined 82–3
  J classes 100–1
  knot clusters 80–1, 84
  Malaysian Grading Rules 94, 98–9, 103
  need for grading 80–1
  NHLA Rules 94–8, 103
  North American softwood appearance
    grading 90–3, 102
  permissible defects 84, 99
  PHND, BHND and sound grades 99
  prime and comsel grades 97–8
  prime, select and standard grades 98–9
  quality defined 82
  Russian softwood qualities 89, 102
  Scandinavian grades 87–9, 93,
    99–100, 102

shipper's usual 99–100
  temperate hardwoods 94–8
  tropical hardwoods 94, 98–9
  unsorted, fifths and sixths 88–9, 93,
    99–100
quality stamps 176–7
Quebec Yellow Pine 29–31
Quercus spp. *see* Oak

radial direction 15
radial movement 8–9, 49
rate of growth 33–8
rays 14–16, 17, 49, 159
relative humidity (RH) 44–5, 50, 63, 194–5
resin 30–1, 84–5
ring-porous timber 25, 34–6, 151, 154
risk 60–1, 65, 78
roofing timbers 64–5, 80–1, 106, 195
Russian softwood qualities 89, 102

sapwood 10–13, 17, 147
sawn timber 4
Scandinavian grades 87–9, 93, 99–100, 102
scientific names 27–31, 37, 139
Scots Pine 13, 140, 142
sea defences 76–7
seasoned wood 47–8, 195
selects 96–7, 98–9, 195
shipper's usual 99–100, 196
shipping dry 39, 52, 55, 56, 62
Shipworm 76–7
Shorea spp. 70, 157
shrinkage 46, 47–8, 62–3, 136–7
silica deposits 12
Sitka Spruce 75, 142
sixths 88–9, 93, 196
size effect 117–18
softboard 181–3
softwoods
  cell structure 23–4
  definition and characteristics 19–23, 196
  earlywood and latewood 24, 32–3, 35
  geographical distribution 21–3, 31–3
  growth rings 31–2, 33–4
  material properties 1–2, 11
  moisture content 48, 54–5
  movement 48
  quality and grading 80, 83, 89–92, 94, 100–2
  rate of growth 33–4, 37–8

species groups 142, 148–9
strength grading 108, 115, 120–1
timber trade 19–27, 31–4
United Kingdom species usage 139–49
use specification 58, 65, 68, 74–7
wood-based sheet materials 166–7, 172–5
sound grade 99
Southern Pine 29–31, 146–7
Span Tables 109–10
special structural (SS) grade 107–8, 112–13
species groups 142, 148–9
specification of moisture content 44–6
splintering 9–10, 17
Spruce 122–3, 141–2
Spruce-Pine-Fir (SPF) 148–9
stains 2, 133–4, 137
standard grades 98–9
steam bending 153
storage 2, 47
straight-grained timber 9–10, 17, 151–2,
    154–5
strength grading 104–21, 197
    appearance versus strength 106–7
    BS EN 1912 standard 108–9, 112
    Europe and North America 112–13,
        119–20
    GS and SS grades 107–8, 112–13
    hardwoods 115–19, 120–1
    machine grading 110–12
    marking systems 119–20, 121
    material properties 5
    SC3, SC4: C16 and C24 grades 109–10,
        120
    size effect 117–18
    softwoods 108, 115, 120–1
    specification of wood species 114–15
    strength classes 111–12, 114–15, 118–19
    TR26 grade 113–14, 121
    United Kingdom 107–11
    visual strength guides 107
stud walls 65–6, 106, 112–13, 197
sustainable forestry 70, 148
Sweet Chestnut 153–4, 159
Swietenia macrophylla 69–70, 156–7

Tanalised timber 128, 197
tangential direction 15
tangential movement 8–9, 49
tannins 12

Tatajuba 161
Teak 161
Teak (*Tectona grandis*) 70, 161
temperate hardwoods
    quality and grading 94–8
    strength grading 116–17, 121
    timber trade 21–3, 31–2, 36–7
    United Kingdom species usage 151–4,
        156–7, 158–9, 162–3
    wood-based sheet materials 168,
        172–3, 175
temperate softwoods 31–2, 36
temperature changes 8
Teredo spp. 76–7
Terminalia ivorensis 155
Thuja plicata 68–9, 91, 145–6
timber framing 65–6, 179
timber trade 18–38
    cell structure 19, 23–7
    earlywood and latewood 24, 32–3, 35
    growth rings 31–2, 33–8
    rate of growth 33–8
    softwoods and hardwoods 19–27, 31–7
    trade and scientific names 27–31, 37
    wood species and trading 19
    wood and timber 18
timescales for drying timber 54–5
toughness 152
TR26 grade 113–14, 121
tracheid cells 23–4, 32–3
trade names 27–31, 37, 139
treated timber specification 129
Triplochiton scleroxylon 159–60
tropical hardwoods
    quality and grading 94, 98–9
    strength grading 116, 121
    timber trade 21–3, 31–2, 36–7
    United Kingdom species usage 154–6, 157,
        160–2
    wood-based sheet materials 168–9, 172–3,
        175–7
tropical softwoods 31–2, 36
trussed rafters 113–14, 121, 198
Tsuga heterophylla 91, 142–3
Tulipwood 163
twisted-grained timber 17

unsorted grades 88–9, 93, 99–100, 198
urea formaldehyde (UF) 171, 198

use specification 57–78
  British and European standards 57–8,
    67–8, 74, 78
  desired service life 69, 73–5, 77–8
  durability and treatability 58–9
  hazard and risk 60–1, 65, 78
  natural durability 58–9, 65, 67–70,
    73, 77
  preservative treatments 123–4, 127–9,
    137
  quality and grading 79
  use classes 59–60, 61–78, 79, 123–4,
    127–8, 137
  wood species differences 58–9
Utile 161–2

vacuum-pressure treatment 126
varnish 131–3
veneers 63, 153, 160–1, 169, 172–3, 199
visual strength guides 107

waisting 76–7
Walnut 162
wane 86, 199
waterlogging 73
weather and boil-proof (WBP) 169–70, 199
weathering 130–1, 137
Western Hemlock 91, 142–3

Western Red Cedar (WRC) 47, 68–9, 91,
  131, 145–6
wet wood 41–4, 49–50
White Oaks 12
wild-grained timber 17
Wood Protection Association 59–60, 75
wood science and timber technology 27
wood-based sheet materials 164–85
  adhesives 171–2
  appearance grading 173–7
  chipboards 177–81, 187
  face quality 172–3
  fibreboards 2, 181–5, 189
  flaxboard and bagasse board 178
  fundamental properties 165–6
  glue bond and WBP 169–70
  hardwoods 153, 157, 168–9, 172–3, 175–7
  marine plywood 171–2
  material properties 2
  oriented strand board 178–81
  outdoor use 170
  particleboards 177–81, 185
  plywood 2, 153, 157, 164–77, 185, 193–4
  softwoods 166–7, 172–5

yacht varnish 131–3
Yellow Pine 28–31, 147, 199
Yew 19